Decision Engineering

Series Editor

Rajkumar Roy, Dean of the School of Mathematics, Computer Science
and Engineering, City University, London, UK

The Decision Engineering series focuses on the foundations and applications of tools and techniques related to decision engineering, and identifies their role in making decisions. The series provides an aid to practising professionals and applied researchers in the development of tools for informed operational and business decision making, within industry, by utilising distributed organisational knowledge. Series topics include:

- Cost Engineering and Estimating,
- Soft Computing Techniques,
- Classical Optimization and Simulation Techniques,
- Micro Knowledge Management (including knowledge capture and reuse, knowledge engineering and business intelligence),
- Collaborative Technology and Concurrent Engineering, and
- Risk Analysis.

Springer welcomes new book ideas from potential authors. If you are interested in writing for the Decision Engineering series please contact: Anthony Doyle (Senior Editor—Engineering, Springer) and Professor Rajkumar Roy (Series Editor) at: anthony.doyle@springer.com or r.roy@city.ac.uk

More information about this series at http://www.springer.com/series/5112

John Stark

Digital Transformation of Industry

Continuing Change

Springer

John Stark
Geneva, Switzerland

ISSN 1619-5736   ISSN 2197-6589   (electronic)
Decision Engineering
ISBN 978-3-030-41003-2   ISBN 978-3-030-41001-8   (eBook)
https://doi.org/10.1007/978-3-030-41001-8

This Springer imprint is published by the registered company Springer Nature Switzerland AG
The registered company address is: Gewerbestrasse 11, 6330 Cham, Switzerland

Contents

Chapter 1
Introduction

The starting point for this book was that I was asked to give a course on Digital Transformation. You might think that shouldn't be difficult in view of all that has been written on the subject. However, most of what I found in articles and blogs was marketing and sales material for particular companies and products. It wasn't suitable for a course aiming to give students, or anyone else, a general introduction to Digital Transformation. As a result, creating the course took longer than expected. I wasn't very happy with this first version of the course, and I redeveloped it twice.

Once the course was running well, the students asked about a course book. Their request led to this book. It provides a general introduction to Digital Transformation, not only for students, but also for anyone else interested in the subject.

The course and the book are based on my experience working with many companies and other organisations. I've found that organisations often start off thinking that Digital Transformation is easy. But they realise later that there's much more to it. I remember one customer, after working on Digital Transformation for a few months, drawing an iceberg and saying, "That's my picture of Digital Transformation now. A small part is visible, most of it is out of sight, lurking dangerously beneath the surface."

There are several factors to be aware of in the submerged part of the iceberg. Firstly, different people in an organisation often have different, even conflicting, understandings of Digital Transformation. Next, people want to transform from the current situation to a future situation. But often they don't have a good understanding of the current situation. And frequently they don't agree on the future situation. Even if they do agree about the target, they find it difficult to describe. Then, they have questions about the technologies that can be used to transform. Similarly, they have questions about what should be transformed. And they have many questions about how to run a Digital Transformation Program in their organisation.

The book aims to answer these questions. It's in 3 main parts. The first part is an introduction. It gives some easy-to-understand examples of Digital Transformation with which most people will be familiar. It describes the technologies of Digital

J. Stark, *Digital Transformation of Industry*, Decision Engineering, https://doi.org/10.1007/978-3-030-41001-8_1

Transformation. It gives definitions of Digital Transformation. It describes the business processes which may be the subject of Digital Transformation. The second part of the book gives examples from several industries. For each industry, a typical starting point is described. Then some examples are given of the way the organisation may digitally transform. The third part of the book describes the activities in a typical Digital Transformation Program. I included this part as, in many organisations, I've found that although people may have great ideas about what to transform, they don't have previous experience of how to go about transforming. They don't know how to transform.

Chapter 2 of the book introduces Digital Transformation by giving some examples of the effect it's had on some everyday activities with products and services. There's nothing revolutionary about these examples but, taken together, they show the extent to which Digital Transformation has already occurred. These examples are visible to everyone. Less visible may be the corresponding changes that the companies and other organisations providing these products and services underwent to make them possible.

Chapter 3 looks at the technologies that enable Digital Transformation. Technologies such as Analytics, apps, Artificial Intelligence, automation, autonomous vehicles, Big Data, blockchain, blogs, Cloud computing, database technology, e-commerce, GPS, Internet of Things, Knowledge Management, machine learning, mobile, robotics, smart connected products, smart phone, social, streaming, telecommunications, the Internet, Virtual Reality, the Web, and websites. The long list of technologies shows why it can be difficult to decide on the best set of technologies to apply in a particular organisation.

Chapter 4 addresses the definition of Digital Transformation. It shows several different definitions of Digital Transformation. These highlight the importance of having a good understanding of several practical examples of Digital Transformation. The lack of a single generally-agreed definition can lead to confusion. The different definitions are often the source of disagreements about Digital Transformation between people in an organisation. Different people read and hear different things about Digital Transformation, and don't have an agreed understanding.

Alone, the technologies that enable Digital Transformation do nothing. They have to be applied somewhere to provide benefits. In a company or other organisation, they're applied in business processes. Chapter 5 addresses business processes. It gives examples of Human Resource, Finance, IT, Quality, Supply Chain Management, Customer Relationship Management, and Product Lifecycle Management processes.

The following eleven chapters give examples of Digital Transformation in organisations in different types of industry. For example, an airport, a passenger transport company, a hospital, a bank, a retail store, and a manufacturing company. The examples are based on many different technologies. They'll help participants in a Digital Transformation Program to get a wider and deeper understanding of the objectives and possibilities of Digital Transformation. The examples describe the current situation in the organisation, the issues and opportunities, the technologies that could help, and the way these technologies may be applied in a particular organisation.

Chapter 17 looks at the differences between the Digital Transformation activities of the organisations described in the previous chapters. It shows there are so many differences that it wouldn't be appropriate to apply the same Digital Transformation approach in all organisations. Instead, each organisation has to find an approach that fits its needs and circumstances. The Digital Transformation of an organisation will affect many of its functions and be expected to lead to benefits in many areas. As a result, it should be addressed with a cross-functional, organisation-wide Digital Transformation Program, run with a Project Management approach. Chapter 18 focuses on Project Management in the context of an organisation's Digital Transformation Program.

Chapter 19 looks at the needs and reasons for involving executives in the Digital Transformation Program. It describes typical executive roles and outlines key executive activities in a Digital Transformation Program. The role of executives in Digital Transformation is an important subject. Many Digital Transformation Programs fail because of lack of executive understanding and involvement.

Chapter 20 gives an introduction to the objectives of a Digital Transformation Program. It describes the need for objectives, and their characteristics. It shows that objectives in different organisations are likely to be different. It gives examples for one particular organisation.

Chapter 21 addresses visioning in the context of a Digital Transformation Program. It describes the need for a vision, a Big Picture of the future, and the expected characteristics of a vision. It positions the vision relative to strategies and plans, and outlines the need for executive involvement in developing the vision.

The subject of Chap. 22 is the governance of a Digital Transformation Program. This chapter outlines the typical role and content of the governance. It describes the leading roles in a representative Digital Transformation Program.

Chapter 23 addresses strategies in the context of a Digital Transformation Program. It gives examples of strategies and of their characteristics. It explains the reasons why it's important to develop a good strategy, and highlights the need for executive involvement.

Chapter 24 looks at implementation planning and schedules in the context of an organisation's Digital Transformation Program.

The subject of Chap. 25 is Organisational Change Management in the context of a Digital Transformation Program. Organisational Change Management is a structured approach, involving many Organisational Change activities, which accompanies and supports an organisation as it proactively changes from its existing organisation structure to a clearly-defined future organisation. The objective of OCM is to successfully achieve this change.

Unless there's a good justification for running a Digital Transformation Program, executives are unlikely to launch it. Chapter 26 looks at different justification techniques such as taking advantage of opportunities, eliminating weak points, and Return on Investment (ROI).

The last chapter, Chap. 27, looks at some other important components of a Digital Transformation Program including a Feasibility Study, the Program Charter, Progress Reports, Status Reviews and Program Closure. It's important that everybody in the Program, especially executives, has a shared understanding of these activities.

Chapter 2
Examples

2.1 Government and Utilities

Let's start with something you may not enjoy intensely. Filling in your annual tax declaration. These days, you may do that online from your desktop computer at home, but a few years ago it was all on paper. And a few years ago you might have paid your taxes by cheque or with banknotes, but now you pay online.

If you need something from your town hall or city hall, you probably don't go there to get it. Most likely the information or the form that you need is on their web site, so you can get it there. And they may even offer e-voting so that you don't have to go out in the snow and queue at a polling station.

Your bills for water, electricity, gas and telephone used to be on paper. Now you receive them online and pay them online. And if you're not happy about something, you can fill in an online complaint form. You don't have to go to the local utility office to complain, or wait for hours in a telephone helpline queue.

A few years ago, somebody used to come and read your utility meter so that your consumption could be calculated. Now that may be done automatically by a smart meter.

2.2 Health

Everything about your health insurance is probably online as well. The insurance company provides a portal where you can see the proposal, the contract, prescriptions, bills and payments. Instead of going to your doctor's clinic, you may ask for an online consultation. You may get an online diagnosis. You may be operated on by an online surgical robot, with your surgeon far away. You may be participating in an online health survey, with a device automatically collecting data from your body and sending it to a hospital. You may have an elderly relative who wears a device that gives medication reminders and, in case of a fall, triggers an alarm and alerts an ambulance.

2.3 Money Transfer

A few years ago, if you needed to transfer money to someone in another country, you'd have gone to a bank, filled in paper forms and waited several days or weeks for the money to get to the recipient. Now you can do it online from home, instantly.

2.4 Education

If you're taking a university or Continuing Education course, you'll have access to the school's portal. All the information you need (apart from the answers to your exams) will be there. You'll be able to see the curriculum. You'll be able to see the dates, times and places of lectures. You'll be able to watch some courses on online videos. Maybe you'll be able to follow some live streamed lectures from home. You'll be expected to hand your homework in online. You'll get feedback from your teacher online. You'll get grades and certificates online.

A few years ago, when students wanted information for their homework, they looked in paper books and journals that were stored in a room called a library. Or they looked in huge paper volumes called encyclopedias which were updated every few years. Now students can effortlessly get an up-to-the-minute version of information online.

If you have young relatives, you may give them electronic learning toys as birthday presents.

2.5 Relaxation

Before relaxing at home after a hard day's work, you may tell your robot vacuum cleaner to clean up the mess from the previous evening. Then, to relax, you may watch a streamed film on your tablet. Or read a book online on your smart phone. Or put your VR headset on and enjoy a virtual reality experience.

One thing you may not do very often is reach for a pad of writing paper, and write a letter to a friend. More likely, you'll send them an e-mail, or chat online. Online, you may also post text, photos and videos for your friends. You may update your website or blog. Online, you may participate in a community. You may like and dislike. You may comment.

You may go online to see an online video of how to assemble some flat-pack furniture you've bought. Or how to install that smart blind you bought. Or how to make a cake. You may even go online and order a 3D print of an object you've designed for your coffee table. You can go online and auction your belongings. You may be an influencer and need to check out updates from your strategic partners. Or you may want to prepare something to share with your followers.

You may play electronic games online. You'll have quite a choice. Single-player, two-player, and multiplayer.

To relax, you may prefer to go to a cinema. An online sales kiosk in the cinema complex will show you details about what's showing and when. It will allow you to select and purchase your seat, and your popcorn.

If you participate in sports, you may wear a fitness tracker that includes sensors to measure and inform you about parameters such as your performance, your location, your heart rate and the number of steps you've taken.

When you're travelling, you may use your smart phone to find the nearest restaurant. Or club. Or museum. You may go to a museum to see an exhibition of Virtual Reality art.

2.6 Ordering and Delivery

If you like reading hardcover paper books, you can go online, choose a book from an online catalogue of books, pay for it online, and have it delivered to your home. You can do the same for other items, such as food, groceries, clothes, a meal, and electronic devices. Alternatively you may select and buy online, and then pick up in a store. If you select home delivery, your package may be delivered by a drone.

You can order a pizza online from home and track its progress from the pizzeria to you.

If the item you want is clothing, you can see, on your tablet, what it will look like on you before you order it. After you've ordered it, you'll receive it at home a few days later. If you don't like it, you send it back.

2.7 Travel

If you need to travel, you can reserve a flight from your desktop computer at home, pay for it online, and have the boarding card delivered to your smartphone. A few years ago, you had to go to a travel agent or an airline office to make the reservation, and receive a paper ticket. And once you got to the airport, you had to queue to get a paper boarding card.

You can reserve and purchase train tickets, coach and event tickets online. You can order a ride, a bike, a car, and a room online.

You may travel in a self-driving car.

A few years ago, you might have struggled with a paper map as you drove to your destination. Now you get online directions and online information about the number of places available in nearby parking garages.

2.8   News, Weather

A few years ago, you might have read a newspaper or watched television for the latest news and weather. Now you get them on your smart phone. You may also use your smart phone to look at maps. And then see the best way to go from A to B on a map on your smart phone, and how long it will take.

You may use your smartphone to find the nearest restaurant when you're away from home. You may use it to find a new friend. You may use it to know when your laundry is ready for pick-up.

Access to your smartphone may be by facial recognition or fingerprint recognition, so you don't need to remember a password.

2.9   Food

There may be an online sales kiosk at the restaurant you choose from your smartphone. It shows you what's available, the full menu as well as today's specials. You use the kiosk to make your choice, order, pay, and get your receipt. Then you pick up your delicious meal at the counter.

An online sales kiosk in a store shows you which breakfast cereals you can choose from, their nutrition details, and today's special discounts. If you're a member of the store's loyalty program, as your enter the store, you receive special offers on your smartphone.

In the office, or at an airport, there may be an online vending machine with snacks and juices. When the machine gets close to empty, it's automatically refilled.

2.10   Home Products

In the past, people didn't communicate with the products in their home. Now, they tell their home products what to do, and get useful information from them. For example, before you leave your office, you can switch on your oven and heating so that everything's ready when you get home. Other products let you know that there's no intruder in your home, and the cat's OK. The fridge orders automatically for you when it sees that stocks are getting low.

2.11   Company Activities

A few years ago, people working in company offices received letters in envelopes from customers and suppliers, and notes and memos on paper from colleagues. They asked their secretaries to type replies.

Now, people in offices get 50 e-mails each day. They type their own replies. They're expected to know how to use many types of software. They're expected to fill in all sorts of forms on their screens. Forms for replies to customers, requests for vacation, details about products, progress on projects, replies to requests to attend meetings, etc. They create parts online. They buy parts online, and pay suppliers online. When they've answered all the mails and filled in all the forms, they make online telephone calls, and participate in online meetings and videoconferences with people in other locations. They may have to leave the meeting early because the sensors on a customer's remotely monitored online product have sensed an unusual condition, and the product has sent a message saying that it needs to be serviced.

2.12   Next

The results of Digital Transformation are all around us. In addition to the examples above you can probably think of many more. The next chapters look at the technology behind these examples, and the changes in companies and other organisations that enable them.

Chapter 3
Technology

3.1 Introduction

The previous chapter gave examples of the results of Digital Transformation. This chapter looks at the technologies that enable Digital Transformation. It's not intended to be a complete history of the development of digital technology. That would take an entire book.

The examples in the previous chapter showed that the changes resulting from Digital Transformation didn't all happen at the same time. That's because the technologies of Digital Transformation didn't all appear and mature at the same time.

3.2 History

Digital technology works using just two states. These may be expressed in different ways. Such as 0 and 1. Or off and on. Or true and false.

The computing and telecommunication technologies behind Digital Transformation have been evolving for a long time. In the early 19th century, Charles Babbage invented the difference engine and the analytical engine, new types of mechanical computing machine.

In the 1830s, Samuel Morse invented the Morse code for electric telegraphy, the sending of coded information in the form of electrical signals through networks of wires. Morse code is digital in that it only uses two states, dot and dash.

George Boole wrote about Boolean logic, digital logic, in the 1840s. Boolean algebra works with two values, true and false.

In 1876, Alexander Bell received a patent for the telephone. Sound waves in a telephone mouthpiece were converted into electrical signals which were transmitted over the wires of a telephone network, then converted back to sound waves in a telephone earpiece. The first telephone exchange was also developed in the 1870s.

© The Editor(s) (if applicable) and The Author(s), under exclusive
license to Springer Nature Switzerland AG 2020
J. Stark, *Digital Transformation of Industry*, Decision Engineering,
https://doi.org/10.1007/978-3-030-41001-8_3

In the 1890s, Guglielmo Marconi developed wireless telegraphy, sending tele-graph signals by radio waves. By turning the radio transmitter on and off for standard durations, messages could be sent in Morse code. In 1901, Marconi transmitted the letter S (... in Morse code) across the Atlantic Ocean from England using a radio transmitter developed by John Fleming. The ... was detected faintly by a receiver in Newfoundland. In 1904, to improve detection of the radio message, Fleming developed the diode (two electrode) vacuum tube. A few years later, Lee De Forest developed a three electrode vacuum tube (triode) which could amplify an incoming signal. Telecommunications took another step forward in 1925 when John Logie Baird demonstrated a television system. Whereas Marconi had transmitted sounds, television transmitted moving images.

The first computers were developed in the 1930s. They used vacuum tubes for calculations. Depending on the input voltage, a triode would either be on (current flowing) or off (no current flowing). Calculations were made by applying appropriate voltages to networks of connected vacuum tubes.

At the end of the 1940s, the transistor was invented. A transistor is digital in that, under certain operating conditions, and depending on the current flowing through it, its output will be either zero (off) or maximum (on). The transistor was much smaller, cheaper and more reliable than a vacuum tube device of equivalent functionality. The transistor replaced the vacuum tube, and led to the growth of the electronics industry, digital computing and digital telecommunications.

The first transistorised digital computer was developed in the mid-1950s. Digital computers superseded analogue computers. They used transistors instead of vacuum tubes for calculations.

The first integrated circuit (or chip) was developed in the late 1950s. A single chip includes many electronic circuits (including many transistors, and many other electronic components such as resistors and capacitors) on a semiconductor base.

3.3   Computer Hardware

Many of the technologies related to Digital Transformation are computer-related. Computers have a central component and peripheral components. The component that runs the programs on your computer, the processor, is part of the central com-ponent. Peripheral components include data input, data output and data storage devices.

In the 1950s and 1960s, computers, referred to as mainframe computers, were housed in several large cabinets, each more than six feet tall. The central component of mainframe computers included the processor, system memory, the memory controller and input/output device controllers.

By the 1970s, minicomputers were available. These were much smaller than main-frame computers. Usually the central component of the minicomputer was housed in just one cabinet, about two feet tall, two feet wide and three feet deep. The peripherals took up slightly more space.

By the 1980s, desktop computers were available. The housing for the central component, the screen, and the keyboard were small enough to fit on a desk.

By the 1990s, the main central components of the computer, the processor, memory and input/output controllers, were available on a single thumbnail-sized chip.

By the 2010s, processors, such as those in your smart phone, were millions of times more powerful than those of the early days of computers. They had much more functionality and were a few hundred thousand times cheaper. Compared to 1970s computers, late-2010s smart phones were about a million times smaller, had about 10 million times more memory, and were about 5000 times lighter.

Various peripheral devices are needed to get programs and data in and out of processors. And to store programs and data. Examples of peripheral devices include paper tape readers, card readers, magnetic tape readers, magnetic disk units, keyboards, touchpads, scanners, printers, USB keys, mice, light pens, paddles, joysticks, microphones, speakers and cameras.

By the 2010s, peripheral devices were many times more powerful than those used in the past. In the late-2010s, USB keys could store 1 TB of data, a million times the storage of 1980 floppy disks. A 1970s keyboard unit was about the size of a bar stool. In the late-2010s, there were keyboards on the screens of smart phones.

Computer users want to see programs and data, so screens were developed to display them. Screens have also evolved through different technologies, sizes and prices. Some 1960s screens used in industry cost $250,000. Similar 2020 devices cost less than $1000.

Computer hardware components have to be laid out and connected, giving rise to the need for Printed Circuit Boards, wires, cables, and plugs.

3.4 Communications and Control Networks Hardware

Another technology related to Digital Transformation that's evolved a lot is communications and control networking. These networks are used for sending something (e.g., voice, data and commands) from one place to another. There are many types of network, depending on what's being sent, what's connected, and the protocols/rules for sending.

You may have a computer network at home or in the office. It may include modems, routers, wires and connectors. You probably also use telephone networks. They may include components such as telephone exchanges, copper wire connections, wireless connections, telephone handsets, and smartphones.

There are industrial process control networks in power plants and companies' factories. They may include, for example, a sensor that reads the temperature of steel in a steel furnace, and sends it over the network to a computer which displays it on an operator's screen. There are financial networks such as SWIFT. There are cable TV networks that transmit TV programs through coaxial cables rather than radio waves. Cable TV started in the 1940s.

Since the 1970s, there's been the Internet, a communications network which underlies e-mail, online chat and the World Wide Web.

And since the 1990s, voice calls have been possible through Internet connections with VOIP (Voice Over Internet Protocol).

3.5 Operating System Software

Computer hardware needs an operating system to tell it how to work.

As computer hardware has evolved, different operating systems, such as OS/360 (1960s), Unix (1970s), Windows (1980s), Linux (1990s), and Android (2000s), have been developed to provide functions and control for computer and communications hardware.

Input/output device drivers have been developed for peripheral devices such as keyboards, graphic screens, mice, card readers and printers.

3.6 Programming Languages

A computer program is a set of instructions to tell the computer what needs to be done to carry out a specific task such as making a scientific calculation, or making a financial calculation, or playing a game.

Different types of tasks need different types of instructions, so different programming languages were developed. In the 1950s, Fortran was developed for scientific computing, COBOL for financial and administrative computing. Later came C (1960s), Java and Python (1990s).

3.7 Database Management Systems

Database management systems are special computer programs that manage all the data being used and produced by other computer programs. The first database management systems were developed in the 1960s.

3.8 Bespoke Programming

In the 1950s and 1960s, organisations hired programmers to develop the programs they needed.

3.9 Computer Graphics

In the 1950s, the input and output of computer programs, typically written in Fortran and Cobol, was alphanumeric. It used letters, numerals, and mathematical symbols. Input was on punched cards and tape, and teletypes. Output was on punched cards and tape, teletypes and printers.

In the 1960s, graphic screens became commercially available, for example, for Computer Aided Design (CAD). Development of video games and video game consoles started. The first computer art was created. Using a computer program, objects could be modelled (created and modified), resulting in a digital representation of the object in the computer. This could then be displayed on a graphics screen.

In the 1960s, the head-mounted display was invented. Development of Virtual Reality software, which simulates a different environment for the viewer, began. By the late 2010s, it was used in many areas including gaming, films, training and design.

The development of Augmented Reality software began in the 1980s. Augmented Reality (AR) is achieved by enhancing the viewer's real-world view with virtual overlays. It's used in many areas including gaming and training.

The 1990s saw the first references to a physical object and its digital twin, its digital representation in the computer.

3.10 Digital Process Control

Use of minicomputers in process control started in the 1960s, for example to implement digitising and servo control of the movable stages of projection-microscope measuring machines [1].

3.11 Enterprise Applications

In the 1960s, some companies started to develop software packages that could be used by many other companies in a particular area of their operations.

Among the first such packages were Material Requirements Planning (MRP) systems. Initially they just addressed the production operations of a manufacturing company. MRP systems evolved into Manufacturing Resource Planning (MRP 2) in the 1980s and Enterprise Resource Planning (ERP) in the 1990s. Depending on the package, they may address areas such as financial accounting, production planning, order processing, purchasing, sales, bill of materials, work orders, scheduling, capacity requirements planning, inventory control, workflow management, quality control, recruitment, payroll, and financial accounting [2].

Customer Relationship Management (CRM) packages became available in the 1990s. Depending on the package, they may address areas such as sales force automation, marketing automation, automated marketing emails, sales promotion analysis, and opportunity management [3].

Supply Chain Management (SCM) packages also became available in the 1990s. Depending on the package, they may address areas such as supply sourcing, negotiation, procurement, order fulfilment, supplier relationship management, and returns management [4].

Product Lifecycle Management (PLM) packages became available in the 2000s. Depending on the package, they may address areas such as Product Data Management (PDM), Engineering Change Management (ECM), Computer Aided Design (CAD), Computer Aided Manufacturing (CAM), Computer Aided Engineering (CAE), Simulation, Application Lifecycle Management (ALM), and Product Portfolio Management (PPM) [5].

Going one step further than a package that could be used in any industry, industry-specific software packages have also been developed. Examples include course management and student management packages in the education industry; and hospital management and patient management packages in the health industry.

3.12   Individual Productivity Software

Some companies developed enterprise applications intended to support the activities of an organisation in a specific area, such as managing customer relationships, or managing the supply chain.

From the 1970s, other companies focused on developing software to increase the productivity of individuals, wherever they were, whoever they were, whatever they were doing. Examples of such software are word processing software, spreadsheet software and presentation software.

3.13   Office, Home, Factory Automation

Office Automation started in the early 1970s. It was based on the application of computers to previously manual office tasks (such as typing, filing, copying and transmitting documents) that made use of typewriters, filing cabinets, photocopiers and internal mail systems.

Home Automation, or Home Computing, began at the end of the 1970s as desktop computers became available.

Factory Automation began earlier than Office Automation and Home Automation. The first Numerically Controlled (NC) machines appeared in factories in the 1950s. A program, rather than a person, controlled the motors that moved the cutting tool on the machine.

The first industrial robot appeared in the early 1960s. A program controlled a handling arm.

3.14 World Wide Web, E-Commerce

In 1991 the first website went live at CERN with the objective of sharing information between scientists. It used the HTTP protocol and the HTML language. These enabled data to be displayed on a web page on a screen, and data to be input on a form on a screen. The number of websites in 2019, almost thirty years later, was estimated at more than 1.5 billion. By that time, the Web was used to share content (numbers, text, photos, videos and audio) between everybody.

By the end of the 2010s, there were many e-commerce web sites that allowed customers to buy online using online purchase forms and online payment. Many of these stored information about a specific type of product or service in a database and then made it available online.

There were sites focused on texts, or on photographs and pictures, or on videos, or on audio (music, voice).

There were sites focused on property for rent and/or sale.

There were sites focused on books for sale or rent.

There were sites focused on products for sale, on spare parts for sale, on services for sale.

There were sites focused on travel (e.g., airplane, coach, bus and train seats; cruise ship cabins; boat trips)

There were sites focused on tickets for concerts, museums, and sporting events.

3.15 Social Technology

Social technologies facilitate interaction between people.

By the end of the 2010s, in addition to e-commerce sites offering products and services, there were many web sites focused on people (e.g., people at home, people in companies, students, drivers of cars, owners of property to rent). Some of these sites allowed people to connect with each other, or create and join groups. Others allowed people to comment online (comment form), or complain online (complaint form). There were sites to send messages, and sites for live chat.

3.16   Embedded Software

In the 1960s, software began to be included in some industrial products such as aircraft, medical equipment, machine tools and robots. It was referred to as embedded software, and ran on "on-board computers". Usually it performed very specific functions, was rarely updated and had limited communication outside the immediate environment of the product.

Embedded software played an important part in automating processes in different industries, for example Automated Teller Machines (ATM) were developed in the finance sector. London Underground's Victoria Line was launched in 1969 with trains driven by an Automatic Train Operation (ATO) system. In manufacturing industry, Factory Automation led to Flexible Manufacturing Systems (FMS), robots, and Automatically Guided Vehicles (AGV).

3.17   Robot Products

In the 1980s "robot consumer products" appeared. They included robot lawnmowers, robotic vacuum cleaners, and window-washing robots.

In addition to their primary functionality, such as vacuuming or mowing, these products had functionality to decide about their situation and actions. This functionality is made possible through the computer hardware and software integrated into the product.

3.18   Related Technologies

In parallel with the computer and communication-related changes, other technologies were also evolving. Xerography started in the late 1940s, document scanning in the 1950s. In the 1960s and 1970s, the Global Positioning System (GPS) emerged. It uses satellites to calculate the position of a GPS receiver. Many smartphones contain a GPS receiver so they're aware of your location. They can provide this information to apps that can look for the nearest cab, restaurant, or new friend.

Radio-frequency identification (RFID) is another location detection technology. It allows products to be tagged with chips that can provide information about the product when it's scanned. This allows products to be tracked throughout their lifetime. RFID offers opportunities to get a better understanding of the way products behave over their lifecycle.

Additive Manufacturing technology started to be developed in the 1980s. Also known as 3D printing, it can be used to rapidly produce a physical part from a digital model.

In the 1970s, credit card systems were computerised. In the 1980s, online banking started. Internet banking appeared in the early 1990s.

Meanwhile, wireless telephone network technology was advancing. 1G was introduced in 1979, 2G in 1991, 3G in 2001, 4G in 2009, and 5G in 2019. From 2G onwards, digital technology was used. In 1992, SMS (short message service) was introduced.

3.19 Mobile Telephony

Before the 1990s, most people only had access to fixed line telephones.

Advances in wireless communications technology meant that, from the 1990s onwards, by using mobile devices such as cell phones, people could communicate from almost anywhere to almost anywhere.

3.20 Related Standards

Standards are needed to facilitate the use of technology. For example, in the 1960s, unique identifiers, International Standard Book Numbers (ISBNs), were introduced for identification of books. In the 1970s, Universal Product Code (UPC) and European Article Numbering (EAN) barcodes came into use. They allowed supermarket checkout scanners to identify the product you're buying, for example 5 011321 833616 or 978-3-319-17439-6. In the 1990s, a standard two-dimensional bar code, the QR code, started to be used.

Internet Protocol version 4 (IPv4) protocol was developed in the 1980s to identify devices on a communication network. Each device had a 32 bit address that could be written in decimal form, for example as 162.23.128.7. However, use of 32 bit addresses limited IPv4 to about 4 billion addresses. More addresses were needed. A successor protocol, IPv6, was agreed. It uses 128-bit addresses.

3.21 Smart Products

The 1990s saw the emergence of "smart products". These products, in addition to their primary functionality, have smart functionality to decide and communicate about their situation or environment. The smart functionality is made possible through the computer hardware and software integrated into the product. Smart products are usually connected to a communications network, which allows them to communicate widely with the world outside the product. This allows the product to report about its condition and environment. It also allows for frequent updates of the product's software. Almost all products can become smart. Examples include smart feeding bottles for babies, smart clothes, smart smoke detectors, smart refrigerators, smart thermometers, smart security cameras, smart pet feeders and smart watches.

3.22 Internet of Things

The notion of the Internet of Things (IoT) emerged in the 1990s. In the IoT, smart products are connected to the Internet. As a result, they're sometimes referred to as smart connected products. The product usually includes several devices: a receiver and transmitter for communication with the Internet; a memory to store data; a processor with operating software; a sensor to sense some characteristics of the product or its environment; an actuator to make the product do something. By the 2010s, advances in computer technology meant that these devices were small enough to fit almost anywhere (e.g., in your smart phone, in your clothes).

3.23 Big Data and Analytics

Big Data is the name given since the 1990s to the enormous volume of data generated in commercial contexts (e.g., Point of Sales terminals), social contexts (e.g., Facebook, search engines) and industrial contexts (e.g., sensors on smart connected products). The huge volume of data is measured in petabytes and exabytes.

The data may be in the form of numbers, texts, photos, videos, etc. It may represent purchases, gender, location, etc. There's so much data that humans can't read it all and make sense of it. As a result, a new technology known as Analytics was developed. Algorithms analyse the data, whatever it represents, and search for correlations, patterns, meaning and other valuable information.

3.24 Packaging Concepts

In the early days of digital technology, computing and communications technology was so voluminous that it took up entire rooms. All of the components of a mainframe "computer" were usually housed together in a special temperature-controlled room.

Gradually, technology became smaller, for example, the IBM PC appeared in 1981. By that time, the "computer" only needed a desktop.

In the 1990s, smartphones were developed. They brought together computer and communication technology in one small device. Miniaturisation of communications and computer devices was mainly due to integrated circuits, chips, getting smaller and smaller.

Also in the 1990s, Cloud computing developed. By this time, all of the components of a "computer" no longer needed to be together in the same place. Some could be on your tablet or smart phone, while others were on faraway cloud servers and other devices that were connected by the Internet.

By the 21st Century, with the Internet of Things, any product (whether a smartphone, an ATM, a sales kiosk, a car or a weighing machine) could include the functionality of a "computer".

3.25   Software Research

From the 1950s, there has been research in software technologies such as Artificial Intelligence, Machine Learning and Knowledge Management. Research in the area of Blockchain technology started in the 1990s.

Artificial Intelligence programs perform tasks normally carried out by humans. They can "think", take decisions and take action. They can be used in areas such as customer service, portfolio optimisation, personalised promotions, autonomous drones, autonomous robots and self-driving vehicles.

Machine Learning is one of the subgroups of Artificial Intelligence. In Machine Learning, a computer learns from datasets, patterns and inference how to carry out a specific task without using explicitly programmed instructions.

Knowledge Management Systems aim to allow the experience and knowledge of humans to be represented and used on a computer so as to increase people's decision-making ability.

A Blockchain is a decentralised networked "ledger" for a community. It records transactions between community members in connected blocks of data. As well as the data about each specific transaction, the block also contains control information such as a date stamp and information on the previous block. Blocks can be added, after verification, to the blockchain, but the blockchain's algorithms and cryptography prevent existing blocks of data (records of transactions) being changed without the involvement of the community. In this way, the blockchain maintains a permanent record of the transaction, and certifies the transaction.

3.26   Next

Huge advances in technology underlie Digital Transformation. There have been huge advances in technology over the last two hundred years. Each advance builds on previous advances. As Isaac Newton wrote to Robert Hooke in February 1675, "If I have seen further it is by standing on ye sholders of Giants." There's no reason to believe that there will not continue to be huge advances in technology.

However, technology alone isn't enough. To produce benefits, technology has to be understood and applied somewhere. The next chapters look at different definitions of Digital Transformation and at the many activities in which the technology can be applied.

References

1. Miller D, Price D, Stark J (1974) The use of a small digital computer in position digitising and servo control of film measuring machines. Nucl Instr Meth 117(2):551–559
2. Monk E, Wagner B (2012) Concepts in enterprise resource planning. Cengage, 978-1111820398
3. Dyché J (2001) The CRM handbook: a business guide to customer relationship management. Addison-Wesley, Boston, ISBN 978-0201730623
4. Hugos M (2018) Essentials of supply chain management. Wiley, Hoboken, ISBN 978-1119461104
5. Stark J (2019) Product lifecycle management (vol 1): 21st century paradigm for product realisation. Springer, Berlin, ISBN 978-3030288631

Chapter 4
Definition

4.1 Customer View of Digital Transformation

One of my customers told me that he was besieged by people wanting to tell him about Digital Transformation. He said that they all told him he needed to digitally transform his company, but from what he said, they all told him different things, and many didn't even know what his company did, or how it was organised. He said that the company had just reported its best-ever financial results, and he wondered what was so bad about that. Why did everyone think he needed to transform?

He said that people were calling and proposing all sorts of things: implementing new applications; changing business processes; training people for new skills; managing change; inventing new business models; applying great new techniques; installing new machines; and adding devices to his products to make them smart connected products.

My customer said that he'd looked at one relatively simple proposal to enable remote monitoring of one of his products, and saw it would affect people in quite a few other parts of the company, such as Marketing, Production, Assembly, Service, Quality, Regulatory, Legal, Patent, Sales and IT. He said 95% of the effort would have nothing to do with digital. And there was no way he could make all those people change the way they worked. He didn't have the authority to make it happen. It was not in his remit. So why did so many people want to tell him about Digital Transformation? And why didn't they agree on a definition of Digital Transformation?

4.2 Examples of Definitions

When I read something about Digital Transformation I always look for a definition of the subject. Sometimes I can't find one. Sometimes there isn't a definition of Digital Transformation.

J. Stark, *Digital Transformation of Industry*, Decision Engineering, https://doi.org/10.1007/978-3-030-41001-8_4

Sometimes there is a definition of Digital Transformation, but it's not a clear definition because it's mixed up with success factors and warnings about Digital Transformation projects.

Sometimes there is a clear definition of Digital Transformation, but it differs from definitions I've seen previously.

As the following examples show, definitions of Digital Transformation abound:

- Digital Transformation applies digital technology to an existing business.
- Digital Transformation aims to create new businesses.
- Digital Transformation is the application of Cloud, Mobile, Social and Analytic technologies.
- Digital Transformation is about changing the company. But it's not just a change due to the application of new digital technologies and services. It's much more than that.
- Digital Transformation is not only a technological change but also as an organisational, cultural and managerial one. Digital Transformation is about reworking strategies, products and processes by leveraging digital technologies. Digital Transformation is really Business Transformation.
- Digital Transformation is the deep transformation of business models and competencies, organisational models, business processes and practices.
- Digital Transformation lets you satisfy your customer who wants, from anywhere, at any time: to get information; to get answers to questions; to buy your product, spare part, or services.
- Digital Transformation improves customer experience through digital technology.
- Digital Transformation is about removing intermediaries from the supply chain, or changing the intermediaries, or becoming an intermediary.
- Digital Transformation is focused on instant availability of information, instant connection, and instant purchase.
- Digital Transformation means transforming the customer's access to shopping and service.
- Digital Transformation is about becoming the fastest (e.g., from a website, app or Big Data) to know what's wanted and then respond fastest to a customer need (e.g., with a book, pizza, or spare part).
- Digital Transformation is the transformation, through the use of digital technology, of our activities.

4.3 Fit of Definitions

The above definitions of Digital Transformation correspond to different viewpoints of the subject. Some definitions are mainly about technology. Some highlight the value for the customer, some the expected value for the organisation. Some focus on the objectives of Digital Transformation, some address the related changes in an organisation.

Clearly, there are many definitions of Digital Transformation. But to what extent do they fit to the examples and technologies outlined in previous chapters?

Having read the above definitions, it's useful to go back and read again the chapter of examples and the chapter of technologies.

One conclusion from the reread is that there have been many different changes in the digital environment. Another conclusion is that there have been several types of digital transformation in different areas. Based on these conclusions, it seems that there could be the need for several definitions of Digital Transformation, with different definitions for the different types of change.

4.4 Types of Digital Transformation

Nine types of digital transformation are addressed in the following sections.

4.4.1 Digital Transformation of Computer Hardware

There have been at least two digital transformations of computer hardware.

The first, in the 1950s, was the transformation from analogue computers to digital computers.

The second transformation was the miniaturisation of computers. The transistorised digital computers of the 1950s were huge. The minicomputers of the 1960s were smaller, but still large. 1980s desktop computers were even smaller than minicomputers. And smart phones in the 2010s are even smaller.

This second transformation has been brought about by use of ever smaller integrated circuits.

4.4.2 Digital Transformation of Software

There's also been a transformation of computer programs. In the 1960s, most organisations hired programmers to write the code for their programs.

By the 1970s, the first software packages were becoming available in a few areas.

By the 1990s, there were many software packages available in many areas.

By the 2010s, there were many industry-specific software packages available in many areas.

4.4.3 Digital Transformation of Communication

The Internet and the World Wide Web have transformed communication.

The Internet underlies chat and e-mails, which have, to a large extent, replaced the sending of letters written on paper by Post Office mail.

The World Wide Web, with its web sites, blogs and Twitter has, to a large extent, replaced communication channels such as newspapers and magazines printed on paper. And it competes with channels such as radio and television.

4.4.4 Digital Transformation of Commerce

Products and services used to be sold from "bricks-and mortar" premises.

Now, many people and companies buy and sell products and services electronically through the World Wide Web and the Internet. For example, through AirBnB, Amazon, eBay, and Uber. Money can be transferred through PayPal.

Monetisation of services may be through online advertising. For example, Google and YouTube.

4.4.5 Digital Transformation of Relationships

Relationships between people usually used to require some physical contact. Or some voice contact by telephone.

Now, many relationships are established electronically, for example, through Facebook, LinkedIn and Tinder.

4.4.6 Digital Transformation of Products

In the first half of the 20th century, products didn't contain electronic and software components. But then, electronic components were added. And then on-board-software. And then there were robot products and smart products. And then products connected to the Internet of Things.

4.4.7 Digital Transformation of Humans

The first artificial pacemaker was implanted in the late 1950s. Since then more and more people have become cyborgs.

4.4.8 Digital Transformation of Society

Together, the digital transformations mentioned above have led to the digital transformation of society in all areas including: government; city planning: military; education; transport; health; financial; retail; publishing; the arts; media; research; leisure; and manufacturing.

4.4.9 Digital Transformation of Industry

The Digital Transformation of industry is the subject of this book. It's addressed in the following chapters.

4.5 Definition of Digital Transformation in Industry

For this book, which addresses the digital transformation of industry, the following definition of digital transformation will be used.

"Digital Transformation is the transformation of part or all of an industrial organisation, through the application of a particular digital technology, or technologies, to improve one or more of its activities".

The "industrial organisation" term is used as, in addition to industrial businesses, there are many other types of organisation in the industrial environment. For example industry associations, regulatory bodies and government agencies such as the Food and Drug Administration (FDA) and the Federal Aviation Administration (FAA).

Chapter 5
Business Processes

5.1 Business Processes

Previous chapters showed some of the new technologies that enable Digital Transformation. Technologies such as Analytics, apps, Artificial Intelligence, automation, autonomous vehicles, Big Data, blockchain, blogs, Cloud computing, database technology, e-commerce, GPS, the Internet of Things, Knowledge Management, machine learning, mobile, robotics, smart connected products, smartphone, social, streaming, telecommunications, the Internet, Virtual Reality, the Web, and websites.

That's a lot of technology! However, technology alone does nothing, it has to be applied somewhere to provide benefits. In a company or other organisation, it's applied in business processes [1].

In the previous chapter, Digital Transformation was defined as the transformation of part or all of an industrial organisation, through the application of a particular digital technology, or technologies, to improve one or more of its activities.

A business process is an organised set of activities, with clearly defined inputs and outputs, which creates business value. Within each of the activities there are tasks that are carried out by people with particular roles. The people may use tools of various sorts to carry out their tasks. The tools can include, but are not limited to, digital technologies.

An organisation's business processes are usually divided into three main groups. These are Management processes, Support processes and Operational processes.

Management processes include Governance, Quality, Financial Planning, and Merger and Acquisition (M&A).

Support processes create value for internal customers, people within the organisation. They include Human Resources (HR), Finance and Administration (F&A), and IT.

Operational processes create value for external customers. The three main operational processes are Supply Chain Management, Customer Relationship Management, and Product Lifecycle Management [2].

J. Stark, *Digital Transformation of Industry*, Decision Engineering,
https://doi.org/10.1007/978-3-030-41001-8_5

Technology can be applied to improve the performance of all these business processes. But, before applying technology in a company or other organisation, it's best to understand the organisation's tasks, activities and processes. That's a first step to improving them.

5.2   Human Resources

Different organisations define their processes differently, but tasks in the HR process often include (in alphabetical order) attending job fairs, employee off-boarding, employee on-boarding, filling job vacancies, hiring a new employee, making payroll adjustments, managing employee status changes, managing executive bonuses, managing new hire requests, managing vacation time and vacation requests, responding to prospective employee enquiries, and updating employee data.

There are many opportunities with Digital Transformation to improve the HR process. Digitally transforming HR tasks could result, for example, in hiring, developing and keeping better people.

5.3   Finance

In the Finance support process, the areas to be reviewed for potential Digital Transformation could include (in alphabetical order) accounting, accounts payable, accounts receivable, approving budgets, capital expense requests, cash management, contract management, expense requests, investment authorisation, invoice reconciliation, invoicing, new product costing, purchase requests, report preparation, and wire transfer requests.

Digitally transforming these tasks could result, for example, in reducing costs and receiving customer payments faster.

5.4   IT

In the IT support process, tasks could include (in alphabetical order) creating, modifying and deleting accounts, delivery and support, handling of incidents, help desk, implementing software, managing licenses, managing permissions, managing requirements, monitoring, planning, reporting, resetting passwords, risk management, security incident reporting, and selecting applications,

Digitally transforming these tasks could result, for example, in providing better service to users and better aligning IT resources with business objectives.

5.5 Quality

The Quality process could include business process management activities such as those for establishing, defining, documenting, publishing, maintaining and improving business processes. In turn, these activities could include tasks for planning, review, measurement, audit, monitoring, verification and validation. Other examples could include developing and maintaining a Quality Manual, training and assisting employees on quality issues, risk management, quality planning, quality assurance, quality control, monitoring and assessing quality, gap analysis, internal audits, external audits, review meetings, training for internal auditors, identifying new processes and developing procedures, establishing appropriate measures and tools, and defining non-conformance procedures.

5.6 Supply Chain Management

Supply Chain Management activities include (in alphabetical order) executing supply chain transactions, handling, inspecting goods, labelling, lot sizing, managing inventory, managing returns, managing supplier relationships, managing work-in-process, negotiation, packaging, physical distribution, procurement, receiving goods, shipping, supplier on-boarding, supply sourcing, transportation, vendor qualification, and warehouse management.

5.7 Customer Relationship Management

Customer Relationship Management activities include (in alphabetical order) analysing sales, creating promotion content, handling complaints, handling queries related to order status, handling returns, managing contact centres, managing sales activities, modifying orders, opening a new customer account, promoting sales, providing customer support, replying to customer messages, reporting sales, running customer satisfaction surveys, taking orders, tracking client history, tracking sales, and uncovering insights and information about potential customers.

5.8 Product Lifecycle Management

The Product Lifecycle Management process often includes six product-related processes. Five of these correspond to the five phases of the product lifecycle. These are the Product Ideation process, the Product Definition process, the Product Realisation process, the Product Support process, and the Product Phase Out process. The sixth process is the Product Portfolio Management process [3].

There's a lot going on in a company as a product is ideated, defined, realised, supported and retired.

The following list shows (in alphabetical order) some of the things that have to happen if everything is to work well with the product across the lifecycle. Analyse Big Data from smart connected products, assemble parts, audit suppliers, capture product ideas, compare actual product costs to planned costs, configure products, contract preparation, contract review, corrective action, cost products, define Bills of Materials (BOMs), define Design Rules, delivery, design control, design products, disassemble products, disposal, document control, equipment purchase, evaluate proposals, get feedback, handling, hire people, identify requirements, inspection, integration, leadership, machine set-up, maintain products, make changes, make parts, manage changes, manage data sent from sensors on products, manage orders, manage partners and alliances, manage projects, measure progress, operations analysis, packaging, part storage, plan manufacturing, plant maintenance, prioritise projects, process control, product modification, progress review, project management, project planning, prototyping, provide service, purchase parts, quality assurance, quality control, recycle parts, refurbish products, replace parts, report progress, retire products, risk management, screen ideas, simulate parts, solve problems, specify products, test parts, train people, upgrade equipment, use products, validation and verification.

The list is far from complete, but it highlights that there are very many activities in the product lifecycle. With so many activities, there are many opportunities for digital transformation [4].

5.9 Industry-Specific Processes

Most of the activities in the previous sections are common to companies in all industries. In addition, a company may have some activities that are typical of its industry. For example, pharmaceutical companies may have industry-specific regulatory, medical assessment and drug approval activities. A bank may have special credit checks for mortgage applicants. A hospital could have special patient medical chart management and pharmacy management activities. A law office could have special activities for restructuring family trusts, or supporting clients in alternative dispute resolution (ADR).

5.10 Mapping the Processes

It's important to be able to identify and list all the tasks in a process. Otherwise some opportunities for Digital Transformation could be overlooked. It's also important to document the processes, to make a detailed diagram that shows the tasks in the process, how they fit together, the roles involved, the information flows, and the tools being used. The activity of documenting the process is referred to as process mapping [5].

Process maps show the organisation's current tasks. They can be annotated with additional information to show, for example, how long a particular task takes. They can also be annotated with additional information to show, for example, the tasks that are at the root of complaints from customers, both internal and external to the organisation.

Process maps demonstrate to an organisation's executives that the organisation's processes are understood in detail. On the basis of the process maps, the company can look for ways in which Digital Transformation can be applied to achieve business objectives such as improved customer delivery or reduced costs.

Creating a detailed process map requires some hard work. However, process mapping isn't rocket science. But it does require a good understanding of the organisation's activities. That's one reason why processes are often mapped by a team, rather than an individual. In most organisations, there are many things going on, and few people know about all the activities. Once the process has been mapped, the process map is shown to other people who work in the process. Often, they'll point out details that have been missed.

As the following example shows, just writing down some of the activities in an organisation to which Digital Transformation could be applied takes time and gets into many details.

5.11 Bookstore Example

A potential customer, living twenty miles from a mall bookstore is looking for a particular book about Antarctica. It's a classic book, but has been out of print for a few years. On Saturday morning, the customer drives to the mall and parks. They walk to the bookstore where they're greeted by the sales assistant.

The potential customer walks round the store, looking to see what's there, looking at the books, looking for the one they want. As they don't see it, they ask the sales assistant, who looks on the store's computer to see if it's in stock. It isn't, so the assistant suggests the customer orders it and picks it up in a few days. The customer agrees. The sales assistant hands the customer a form, which they fill in with the book title and author. The sales assistant enters the information in the computer.

At the end of the day, the bookstore owner prints out a list of books ordered that day. After walking round the store, just to make sure the books really aren't in stock, they call a few warehouses to see who can offer the best deal on delivery time and cost. Once they've found the best deal, the bookstore owner orders the Antarctica book.

In the warehouse, the book is located, packaged and weighed. The shipping assistant calculates the postage, and affixes the address label and the stamp. Late that afternoon, all the books ordered from the warehouse that day are taken down to the Post Office.

Two days later, the package with the book about Antarctica arrives at the mall bookstore. The bookstore owner unpacks the package, checks it's the right book, and puts it on the shelf reserved for ordered books.

On Saturday, the customer drives to the mall and parks. They walk to the bookstore where they're greeted by the sales assistant, who gets the book from the shelf and hands it to the customer. Great, that's the one! The sales assistant asks the customer how they want to pay for the book. It's $20. Cash please! The customer hands over a $20 bill. The sales assistant rings up the sale, puts the money in the till and asks if the customer would like the book wrapped. Yes please, it's a birthday present for my brother!

Later that day, the bookstore owner puts the day's takings in a bag, and deposits it at the bank in the mall.

5.12   Importance of Mapping

The above description of events around the sale of the book may seem unnecessary, too detailed, overlong, boring and tedious. However if you want to apply Digital Transformation, you have to understand in detail from where you're starting. In any organisation, there's a lot going on, and it's easy to forget something. But if you forget something, you may miss an opportunity. That's why it's important to document the business processes.

5.13   Mapping

You can try to map the process for the above example, starting by listing the roles of the people, and noting the main activities, for example, ordering, invoicing and shipping. The roles include customer, bookstore owner, bookstore assistant, warehouse salesperson, warehouse packer, warehouse shipper, post office worker, bank worker, and customer's brother. You can also note the computer systems, such as the bookstore computer, the warehouse computer, post office computer systems, and bank computer systems. And then the flow of tasks. And the flow of information.

The above example is just one of many for a bookstore. There's another example in which the customer finds the book when they first go to the bookstore. The details of some of the tasks will be similar, but there will be differences in their order. There are also other activities to consider. What happens if it turns out that the customer's brother already has this book on Antarctica, and the customer asks to exchange it for a book about polar bears? And what happens if the customer's brother starts reading the book, and complains that Page 45 is missing?

And in the background, there are other bookstore processes such as human resource (e.g., hiring, payroll) and financial (e.g., accounting).

5.14   Moving Forward

Even in a relatively simple organisation such as that of a mall bookstore, there are many ways in which Digital Transformation can be applied to activities.

The following chapters look at ways in which Digital Transformation can be applied in different types of organisation.

References

1. Dentch M (2016) The iso 9001:2015 implementation handbook: using the process approach to build a quality management system. ASQ Quality Press, Milwaukee. ISBN 978-0873899383
2. Stark J (2019) Product lifecycle management (vol 1): 21st century paradigm for product realisation. Springer, Berlin. ISBN 978-3030288631
3. Stark J (2016) Product lifecycle management (vol 2): the devil is in the details. Springer, Berlin. ISBN 978-3319244341
4. Karniel A (2011) Managing the dynamics of new product development processes: a new product lifecycle management paradigm. Springer, Berlin. ISBN 978-0857295699
5. Madison D (2005) Process mapping, process improvement and process management. Paton Press, Chico. ISBN 978-1932828047

Chapter 6
Digital Transformation at Springfield Council

6.1 Introduction

This is the first of several chapters looking at Digital Transformation in different sectors of the economy.

6.2 Springfield Council

Springfield Council is the legislative branch of government for Springfield. The council provides a wide variety of services to its community. It is active in areas such as: planning; building; education services; health services; introducing local laws; law enforcement; community services; waste management; animal management; recreation and culture; traffic, roads and parking; and emergency management.

6.3 Current Situation

Springfield Council built their own website. They're also on Facebook and Twitter. Their website has a mass of information. It includes:

- a description of products and services provided by the council;
- administrative information such as tax rates, tax payment dates; administrative forms for download (birth, marriage, death, new arrival, departing inhabitant, animal licences, entertainment licences, street market licences, etc.);
- videos showing how to fill in forms;
- a list of polling stations;
- a list of elected officials;
- a table of election results;

J. Stark, *Digital Transformation of Industry*, Decision Engineering, https://doi.org/10.1007/978-3-030-41001-8_6

- a list of dates of council meetings;
- a diagram of the council's organisational structure showing contact information;
- a subscription form for the weekly e-newsletter "What's happening in Springfield";
- a list of government buildings and offices with opening times;
- a list of community centres;
- a feedback form for comments about the site;
- a section about the history of Springfield;
- podcasts from the Mayor;
- an order form for online purchases;
- a query form in case inhabitants have a question;
- answers to Frequently Asked Questions;
- information about price reductions available to inhabitants;
- maps; a list of street names and geographical feature names;
- utility maps; planning information and applications;
- information about adult education centres and courses;
- information about winter road gritting;
- details of Park and Ride;
- disabled parking information;
- a list of recycling sites;
- advice on trading standards;
- youth clubs and children's play sessions;
- details of parks and green spaces;
- annual public holidays and other events;
- job vacancies;
- press releases;
- Latest News.

In a password-protected area, there's:

- a list of voters;
- a list of inhabitants;
- a list of council employees.

There are links to other local organisations (clubs, sports grounds, shops, cinemas, theatres, hospitals, doctors, dentists, jails, police, fire service, buses, trains, airport, bicycles for rent, schools, restaurants, car parks, refuse collection, swimming pools, libraries). There's a link to weather and flood warnings.

6.4   Looking Forward

Springfield council is working on developing a digital transformation strategy. Its objectives are to reduce costs and improve resident experience. One possibility being investigated is to develop apps to facilitate payment of local services, support ride sharing, and show currently available parking spaces. Other possibilities being investigated include:

- adjusting traffic lights to traffic flow;
- switching street lighting on and off depending on actual conditions, not pre-set times;
- monitoring noise and other pollution;
- increasing use of smart energy meters to reduce energy usage;
- introducing instant fines for road speed and parking violators;
- providing electric car recharging stations;
- moving to energy efficient buildings;
- supporting the circular economy;
- building an incubator and eco-system for start-ups;
- offering courses in Digital Transformation.

The council is also rethinking its current policy of providing free links to other local organisations. It's looking at ways, such as affiliate marketing, banner advertising, and publishing sponsored content, to monetise its services.

6.5 Looking Outward

As part of its efforts to develop a digital transformation strategy, Springfield council members visited other councils to see what they are doing in this area. The other councils were found to have taken several different approaches. Many had developed plans to enable use of digital technology to deliver better services and to improve the customer experience, yet save money. Some had introduced new technologies such as route planning tools and mobile telephony to make their workforces more productive, and reduce costs. Some councils offered free wireless access in their buildings. Many had put applications and data in the Cloud. To deliver financial savings, some had worked together to reduce the cost of technology. Some councils had created separate websites for different publics, for example, a family website, a business website, and a carer website. In general, technology was being seen as a way to radically reshape the way services were delivered. Springfield council learned of many ways in which it could improve its digital offering, for example:

- using tele-care and other assistive technologies that enable the elderly, sick and disabled to continue living at home rather than in a hospital or a council-run retirement facility;
- offering online paper-less application for school places;
- enabling cashless parking payment;
- using GPS technology to improve the quality yet reduce the cost of waste collection and cleaning;
- installing self-service kiosks for members of the public to self-serve for services such as requesting a bulk waste item collection, making a payment, and reporting environmental problems such as graffiti and fly-tipping;
- providing a smart phone app enabling relatives and neighbours to keep in touch with the lonely and socially isolated;

- responding to questions through web chat rather than by phone call;
- analysing socio-economic data to better target services on those who most need them;
- installing smart sensors in parking bays and on road surfaces.

During discussions with other councils, one council suggested Springfield Council ask townspeople for their view of its e-services. Unfortunately, some townsfolk told of bad experiences. One person said that it was written on a Council web page that they would be mailed two weeks before a payment was due. They never received a mail. And were then made to pay a surcharge for not paying on time. A second person said their credit card information had been stolen. One woman gave an account of the Council website proposing chat assistance. But the Digital Assistant didn't seem to understand a simple question, and for some reason kept repeating itself. Several people pointed to Council webpages where the information wasn't up-to-date and correct. Someone pointed out a Council webpage with a recommendation and opening hours for a restaurant that no longer existed. Others mentioned getting "404: This page doesn't exist" messages. Many reported filling in forms on the Council website, clicking "Send", and never getting a reply. Some complained their input to Council forms had been rejected without an explanation. Another individual stated that when they tried to make a payment from a Council webpage the payment was rejected without reason. One man recounted a story of signing up to Council services with a username and password. But when the man tried to use the service a few months later the username and password kept getting refused. The man then called the help number, and after about twenty minutes someone replied. After the man had told his story, the help person had said they now used two-step verification so the man should choose where he wanted to receive the password then click "Enter". And hung up. The man said he didn't understand this, so drove to the Council offices to talk face-to-face with a human and ask what's occurring. But the Council offices were already shut for the day. One company said the confidential information they had provided when registering as a new supplier had been shown to their competitor. Various people complained about the Council website being too slow, and too hard to read. Others said the design was outdated, or over-cluttered, or they couldn't find contact details. A farmer said he sent the Council a request for a fishing permit and made the payment. He said he got an e-mail back saying he could follow his permit request's progress with a 6 number tracking number. He looked on the Council site, entered the number and saw his request was in Received status. A week later the permit arrived at his farm. He looked on the Council side, entered the number and saw his request was still in Received status. He said that having a tracking number looks great, but shouldn't the status have been updated to Sent or Completed or Delivered? A local resident complained that he couldn't enter his name on a form on the Council website. Every time he tried, he got an "Illegal name or character" message. He said he didn't understand because he often got bills, in his name, from the Council. He said it wasn't right that his name was OK when the Council wanted his money, but not when he asked for a Council service. An ex-council member complained that they had lost in an election because there was a software issue with a smartphone app. A householder complained that the Council's website was sexist and ageist. About three-quarters of the photos on the website were of females under 30 years, but 50%

of the population was over 50, and more than 50% was male. A partially sighted person said they couldn't use the Council's website. It was only AA level accessible. They hoped the Council would soon be Triple A compliant. Somebody asked that the Council stop changing the layout on the screen as that made it difficult to find what they were looking for. A man suggested there should be a way to remove the chat panel as it took up a lot of room. A dog owner claimed that changing the website was just covering over the cracks, the real problems with the Council were much deeper. A person with a disability said they had to click through too many pages, so hoped for more text but fewer pictures on each page. A doctor complained that the text on some pages was too small, but when they did "spread fingers" the text didn't get bigger. An elderly person complained that for a year the "hot meal" they received each day from the Council was always cold. Then they were told it was because there was a coding error. Nobody apologised. It seemed nobody cared. From its discussions with other councils, Springfield council also learned that there are several cultural issues around Digital Transformation. For example, those of sharing personal data, and of trusting government organisations. Many members of the public were found to be concerned about correct and secure use of their confidential personal data. Other residents felt that digital transformation was a disguised way to reduce their benefits and services. Some felt left out because they were unable to use digital technology. Many council employees saw transformation as a way to eliminate their jobs. The council realised that it would need to put in place a change management program to help employees and residents adapt to the new digital environment. Digital Transformation is all about change. It requires people to do things differently. Some people will welcome change, while others will resist. The council wants to be sure that people will understand both the need for the changes and the changes. It wants to be clear about communicating what people may lose but, equally, to show the benefits they will receive. People need to be aware of the council's transformation journey. That requires reaching out and building awareness, addressing issues, and gaining the trust and support of the community. Springfield council also learned other lessons from its discussions with other councils. These lessons included: people, activities and digital technology are closely entwined and need to be addressed together with a holistic approach, otherwise the expected transformation won't materialise; strong leadership from the top is essential for Digital Transformation; keep away from the technobabble and the buzzwords such as Smart City. They may sound great, but first you need to understand and meet your customers' current everyday needs and expectations; collaborative procurement with other councils is one way to reduce costs; communication channels with other councils are important for sharing experience and ideas; one of the big challenges for Digital Transformation is to find people with the right skills. Collaboration with other councils is one way to address this and acquire the needed specialist skills.Another lesson is that before starting detailed transformation activities, it's important to be clear about: Digital Transformation objectives and strategy; the foundations of the Digital Transformation Program such as Program Governance, Program Management and Program Performance reporting; the Program plan and the corresponding detailed business case for change; and the availability of digital skills.

Chapter 7
DT at Springfield State University Institute

7.1 Introduction

Springfield State University Institute (SSUI) is a leading educational organisation. The Institute's mission is to contribute to society through education and research. It aims to educate the next generation, preserve existing knowledge, discover new knowledge and disseminate knowledge for the benefit of all. The Institute offers high quality academic and professional programs at undergraduate and postgraduate levels.

7.2 Background

Last year we worked with a leading global consulting company to prepare our digital future. As a first step, we carried out a current-state assessment. This showed that our structures and processes have changed little in recent decades in spite of many changes in the world around us.

Our consultants found that, historically, our operating model was based on four main activities: doing research to create knowledge; storing knowledge in the library; transmitting knowledge to the next generation (teaching); and certifying by examination that students had reached a required level.

In recent years, our role in storing knowledge has been reduced as so much information has gone online. The Web offers much more knowledge than our library could hold. Google provides instant access to this knowledge. The number of librarians has declined. However, the number of administrative staff has greatly increased.

Our principal delivery model for transmitting knowledge was the onsite classroom lecture model, in which one of our professors, or one of their graduate students, read their notes to a class of over a hundred students. This model had been tried and tested over the last 900 years since the establishment of Bologna and Oxford. Students took

J. Stark, *Digital Transformation of Industry*, Decision Engineering, https://doi.org/10.1007/978-3-030-41001-8_7

notes as the lecturer spoke, then repeatedly read their notes to memorise the content and regurgitate it in examinations.

In recent years, the profile of our teaching staff has changed significantly. They used to be nearly all tenured staff with full-time permanent contracts. Now, about half are adjuncts with short-term temporary contracts.

Most of our tenured academic staff have taken the tried and tested career path from bachelor degree, to master degree, and doctorate, then become research assistant, research associate, lecturer, Assistant Professor, Associate Professor, and finally Professor. Our consultants noted that this path has resulted in many academic staff having no experience of the outside world. While that may not have mattered when teaching Latin and Greek in the 17th century, it can be a disadvantage in these fast-changing times. For example, our consultants found that the knowledge of our academic staff often lags behind that of people working in local companies. In many of these companies, there's a feeling that, because they've been forced to respond to change, or go out of business, their knowledge and know-how is now years ahead of that of our staff who haven't been under such pressure. Some local companies have even set up their own Academies, saying that they can learn nothing from our academic staff. They say we are out of touch with the needs of today's world.

Many of our academic staff are, of course, great Subject Matter Experts. However, some have failed to be continuous learners, and have not kept up with the fast-changing world. Also, some may lack professional teaching skills, such as presentation skills, and skills to engage with students. Others have poor communication techniques, poor interaction with students, and are not seen as skilled educators. With tenured staff, it's difficult to bring about change. Change can always rebound on those asking for change, so there's a certain *corvus oculum corvi non eruit*.

Our consultants found out that new methods of learning have been proposed in recent years. For example, Piaget's Theory of Cognitive Development. They suggested our staff could consider these.

To ensure "academic freedom", our consultants found that we have exercised little control over our academic staff. Every few years, they have a year off, but there is no control of their activities at this time. We don't know to what extent they have developed. And each year, many academic staff members take long vacations, usually during a 3 month period when there are no students on campus. Again, there is no control of their activities at this time. We don't know to what extent they use this time to develop professionally.

In the past, we had a once-in-a-lifetime teaching environment. We taught the student for a few years. Then the student left, and usually never returned. The students were happy with that, they didn't complain. But students have changed. Now many lack self-discipline, binge-watching snowflakes with their avocado toast and selfies.

In today's new world, our consultants see a need for a lifelong learning environment in which first each student learns to learn, and then returns several times to the university to continue learning and developing their knowledge assets. We want to offer the opportunity for each student to be able to learn what they want to learn in their own area, when they want to, at their own speed, in their own way.

Our consultants found out that we are seen as not being focused on the needs of our students, but on those of our academic staff. Our performance-measurement model has been based on the number of research papers that our academics write, and their participation on boards of academic conferences and academic journals. Our performance-measurement model has not been related to what our students, our "customers", to use consultant-speak, do after they leave us. We have no system to follow up. We don't know if they've found a good job they're happy with. We don't know how much debt they leave with. We don't actively maintain contact although of course, if we find out later that they have a well-paid job we ask them to join an alumni group and fund our activities.

Our consultants also found that we have paid little attention to our "product and service portfolio" in the past. In many ways, we have not managed it at all. There has been no review process. Typically it has taken years to discontinue a course, however poor. The consulting company showed that it usually takes us three to five years to enter a new subject area. And at least nine months to add a new course to the curriculum in an existing subject area. We have not been able to offer course options as these would have been too complex to manage. Although we teach courses about innovation, when it comes to applying this knowledge to ourselves we seem to have been remiss.

Working with the global consulting company, after extensive survey, research, analysis, and consultation we were led to see that digital transformation is necessary. We concluded that we can't keep going the old way. We have to adapt, or go the way of the dinosaur. *Tà pánta rheî kaì oudèn ménei.* Fortunately for us, education is becoming the foundation of our society and our economy. Our consultants warned us that one of the main barriers to digital transformation is resistance to change. They said that digital transformation requires non-negligible behavioural adaptation. People across the campus will need to learn new facts, think differently, be open to new ways of working, learn how to use new tools, forget old procedures and adjust. The consultants said that to succeed with Digital Transformation everyone on campus has to change their mind-set. We can't expect the IS Department to do Digital Transformation for us. As the members of the organisation, we all need to find out how to integrate digital technology into our activities in ways that suit our circumstances. Our consultants suggested that, based on their deep experience with similar organisations, they would help us with our transformation. We agreed. We are crossing the bridge to Digital Transformation. *Iacta alea est.* Our consultants will develop a communication program explaining why it's important to transform, how we will transform, and the benefits of digital transformation that we will achieve. The communication program will include workshops, a digital transformation project blog, intranet updates, poster displays, town hall meetings, lunch-and-learn sessions, and after-works. Our consultants will help us establish networks within the organisation to share digital expertise. They'll help us change the remit of some departments. They'll make sure we include digital in all discussions. Our consultants will help us identify and train so-called change agents. These are people working on the campus who will help change things for the better. Some are faculty, some are students. Some are from departments such as Facility Management and Finance. The consultants are also guiding the members of our leadership team, many of whom are digitally illiterate, poor managers of their own time, and unused to delivering anything outside their area of expertise. Our consultants have said that it is truly important for our leadership team to get involved with and guide the transformation. And to set the example for everyone, and to be open to change and open to feedback. As our consultants told us, if high-visibility senior leaders don't adapt their way of doing things, it's unlikely that anyone else will change their habits. As part of the leadership team's activities, the consultants are organising a series of workshops in areas such as leadership development, pedagogical methods, professional development, recruitment, performance management, collaborative and interdisciplinary skills, and technical training. There's been a high degree of satisfaction with these workshops. Among recent workshop results have been suggestions for systems that automatically: provide faculty with detailed information about the way each student is progressing, their most recent progress metrics, and academic advisories such as their potential career paths; provide students with individualised work-plans, taking account of their current progress, their measured learning ability, and results of 24/7 surveillance tracking of their whereabouts; identify alumni to target for specific fundraising initiatives; grade a student's continuous assessment material and examination answers, checking for plagiarism and other forms of academic misconduct. Many issues have arisen, for example, the need to address: personalisation of education; more immersive learning experiences; competition from the rise of non-traditional channels that claim to offer learning; students from industry who request education in line with demands from their employers, and who want a modular approach to learning; faculty interactions with students from industry and commerce who request access outside traditional academic office hours; course creation for virtual learning environment (VLE) systems; and development of a student's special skills and abilities to extend their creativity and innovative potential.

7.3   Our Digital Journey

We have started our digital journey. However this is not an easy road to take. We are aware of many barriers to progress as we move forward.

For example, our structure and culture make it difficult to change. We have many departments and they are all independent and have their own ideas. All departments are considered equal, none is *primus inter pares*. Rather it's *aequales omnes*. There's much inter-departmental rivalry, and inter-departmental disputes are frequent. Many department leaders are looking forward to retiring in a few years, and see no benefit in change.

Another barrier to change is that digital technology has a high initial cost, and a concomitantly high maintenance cost for future years. We lack digital technology capabilities for today's world, and correcting that will have a high cost. And we lack people with the skills in new subjects needed for the future world. Historically our academics have not ventured into new areas but have retrod the same path they took in previous years.

Our consultants found that our Information Technology resources have developed in a disjointed fragmented way. For example, first installing a financial management system, then a human resource management system, then a staff management system, then a course management system and then a student management system. We have selected systems on the basis of lowest cost, with the result that we have many unconnected systems from many system suppliers. We have tended not to make training obligatory, with the result that many people are unable to use the systems.

Our consultants found that our IT staff has expanded steadily in recent years, but without any strategic direction. Our IT staff created our web site and profiles on social media. The web site has sections for potential students, actual students, academic staff, other staff and alumni. There are sections describing the University and the local community. There are sections describing our research and our courses. There is a News section run by our Press Officer, who also writes our Twitter feed.

7.4   IT Staff Dissatisfaction

Our consultants noted some dissatisfaction among IT staff members, many of whom have been with us for a long time.

According to our consultants, the culture of our organisation has changed. There used to be a culture where people could join the university and work for it all their working life. Over time, they understood more and more about the university and its activities and what they had to do. They became more senior and moved up the hierarchy. They concentrated on their job, and got to understand the university's way of working and its rules.

But now a lot of young people are being hired into IT, and they don't even expect to stay more than a few years with the university. They come in well aware of new technologies, and expect to find them in use in the university, with everything happening quickly. They think that a lot of the university's rules, such as badging in and out, don't make sense. They question why HR doesn't have rules against all the sexist behaviour, rather than about the time spent in the restroom. Instead of listening to and learning from the seniors, it's said they're always on social media texting their friends, or following celebrities and other influencers. They don't seem interested in what's happening in the university, but know all the latest fitness, fashion, beauty, car, cooking, and health food tips. In meetings they don't let superiors talk first, just blurt out their ideas. They don't like the university's hierarchical structure, saying it's too slow, and should be changed to be flat and flexible. They talk about people needing to share knowledge, not hide and hoard it. They expect to get information immediately, just like from their smart phone, not wait a month to get it in a faculty meeting.

The IT people who've been in the university for decades tell them they need to take the time to understand the university, not just talk about new technology and unicorns. After a few months, the new hires tend to burn out, apparently frustrated by the environment.

7.5 Digital Transformation Strategy

We defined our Digital Transformation as the full, integrated application of digital technology throughout all operations of Springfield State University Institute. With our consultants, we developed a Digital Strategy to integrate Digital Transformation into a new business model for SSUI. We developed a corresponding Strategic Digital Plan.

We will develop new student-centred ways of working in order to continue to deliver to our mission of providing a high quality learning experience in the face of changing technological, competitive and customer requirements.

We will extend our revenue streams to embrace government-funded teaching grants, government-funded research grants, student fees, research projects with government agencies, borrowing, charities, endowments, philanthropy, naming opportunities, as well as educational and research partnerships with local companies.

We'll connect our products and service portfolio to our strategy. We'll grow our portfolio of products and services, developing new courses and new fields of study to reflect changes in student needs as well as advances in knowledge.

We'll attract the best students, building on world-class outreach activities. We'll increase student numbers. We'll increase numbers in STEM subjects, and increase female and other minority group representation.

We'll digitally transform our student's learning experience with the use of Fourth Industrial Revolution technologies such as artificial intelligence, blockchain, deep learning, real time simulation, big data analytics, gamification, immersive technology, virtual/augmented reality, machine learning, robots, serious games, and courses delivered by 3D holographic professors.

We'll set up a Digital Transformation team including administrators, academics and students to ensure we understand and continue to understand digital literacy needs, and educate at the corresponding level.

We'll grow our national and international business by developing an online learning channel of education.

We'll deliver an unmatched technological experience through our highly skilled IT group. We'll transform our IT organisation from a cost centre to a value adding team. Its members will help us use IT to enhance teaching and learning, and to extend our research capabilities.

7.6 Going Online

Our investigation of e-learning, online learning platforms, virtual learning environments and MOOCs has highlighted the potential of off-site, remote learning. We found the remote learning approach to be easy to use, and available on a 365/24 basis. We found it enables self-paced progress for students with diverse possibilities to rerun lectures. Students can participate from anywhere at any time. The number of students in a class is unlimited. And we found that we can offer blended learning models that integrate online learning with classroom lectures.

As a result of our investigation, we'll develop an online open-source learning platform, offering features for course development, course delivery, classwork, grading, and homework. The platform will include functionality such as collaborative virtual classrooms and course management. Our academic staff will be able to give courses to students throughout the world.

Through this online platform, we'll offer a full online curriculum with: high-quality courses; video lectures and live webinars with our departmental subject matter experts; peer discussions among students moderated by our experts; real-life case studies; challenging assignments; quizzes and interactive assessment. Students will be able to carry out real-life projects addressing challenges in their organisations.

We'll partner with local companies to provide internal training to their employees. We'll offer them in-workplace curricula addressing both hard and soft skills, and supporting lifelong learning for their employees. We'll partner with leading educational organisations in other countries to offer them remote learning and overcome their issues with geographical difficulties, lack of teachers and overfull schools and classrooms.

7.7 First Digital Step

Our first step towards digital transformation will be a transitional step.

We'll choose a cloud-based platform providing free and open source office productivity applications including a word processor, a spreadsheet application, a multimedia presentation application, and a database tool. We'll move our administrative systems to the Cloud to deliver efficiencies. We'll deploy a document management system for storing, retrieving and reporting administrative information, digitise administrative content, and automate administrative processes. We'll implement a new Cloud-based payment system.

Also in our first step we'll review our use of social media. We'll deploy a cloud-based ad management system to better target our customers. We need to focus advertising for Bachelor degrees on the 15–18 year category, and advertising for continuing education on those over 25.

We'll also redesign our web site and develop new mobile apps.

7.8   Second Step

In our second step towards digital transformation, each student will be able to access any information and tools they need through SMEP (Springfield's My Education Portal). This could include their classes, class schedules, coursework, administrative information, and the online library, as well as productivity and collaboration tools.

SMEP will have a user-centred interface. It will include a Digital Assistant that will answer any questions the student may have about student life at SSUI.

In the second step, all students will be required to be equipped with a computer, tablet or smartphone.

Also in the second step, we'll train everybody to use the new tools.

And academic staff will be required to take Digital Transformation courses. By this time, we'll be offering a complete spectrum of Digital Transformation courses. We'll start in the next academic year with the following five Digital Transformation courses.

1. Online Digital Certificate course in Digital Transformation, $2,500. This is a 6 week course, 4–6 hours per week. It counts as 4 continuing education units. Apart from the usual video lectures, the online course includes case studies, demonstrations, interactive activities, peer discussions, challenge projects, assessments, and live Q&A sessions with our experts.
2. One-week Certificate course in Digital Transformation, $9,500. This is an on-campus. It runs Monday to Friday, 5 days.
3. Digital Transformation Diploma, $28,500. This modular 15 day on-campus course is made up of 3 week-long modules. Each of these runs Monday to Friday.
4. Digital Transformation Master, $62,500. This is a 12 month, full-time, on-campus course.
5. Executive MBA in Digital Transformation, $85,000. This is an 18 month, part-time, for working professionals including C-level executives, Business Unit Managers, and Departmental Directors. The course incorporates 1 week study trips to Silicon Valley, Zhongguancun, Bengaluru, and London.

Each study trip will include: discussions with senior Digital executives; excursions to leading Digital companies; meetings with Digital Transformation Program Leaders; and forums with Venture Capitalists. The course also takes in 1 week per quarter working on campus with top industry and academic Digital Transformation thought and practice leaders. There'll be boot-camps, ice-breakers and get-togethers. Each participant will lead a real-life Digital Transformation project. In addition to the on-campus and on-site activities, the course includes 10 weeks of online courses, 4–6 hours per week. Course modules include:

(a) Introduction, Background and Overview of Digital Transformation;
(b) Opportunities and Benefits of Digital Transformation;
(c) The Role of Data in Digital Transformation;
(d) Business Processes in Digital Transformation;

(e) People in Digital Transformation, New and Changing Roles. Digital HR. Digital Mind-sets. Change Management;

(f) Technology in Digital Transformation: Analytics/Big Data; Artificial Intelligence; AR/VR; Blockchain; Cloud; Cybersecurity; Industry 4.0; the Internet; the Internet of Things; the World Wide Web; etc.;

(g) Digital Business Models. Digital Ecosystems;

(h) Digital Marketing;

(i) Industry examples from Advertising, Banking, Government, Healthcare, Retail and Transportation;

(j) Examples of Digital Disruptors. AirBnB, Alibaba, Amazon, eBay, Facebook, Google, Instagram, LinkedIn, Netflix, PayPal, Pinterest, Skype, Spotify, Tencent, Tinder, Twitter, Uber, and Wikipedia;

(k) Innovation Incubators. From Start-up to Unicorn;

(l) Your Digital Transformation Program: Phases; Steps; Roles; Tools; Critical Success Factors;

(m) Pitfalls of Digital Transformation Programs. Overcoming Organisational Obstacles: Communications; Training; Reward Systems; Role Definition;

(n) Leading Digital Transformation;

(o) Your Strategic Action Plan.

7.9 Third Step

In our third step towards digital transformation, our target is that less than half of courses will be held in lecture theatres. Most courses will be in online collaborative workspaces. At this time we also expect to be starting to replace real-life teachers by robotic Digital Teaching Assistants and holographic professors.

7.10 Fourth Step

Having transformed the approach to classroom lectures, in this step we'll start to address practical, hands-on, "lab" work.

Chapter 8
Digital Transformation at an Art Collective

8.1 Introduction

Creative Artists Working Collective (CAWC) is a community of artists, working in the buildings of an abandoned farm, hundreds of miles from the nearest large city. CAWC used to work with a city dealer who would display and sell their paintings, and take a percentage of the sales price.

8.2 Current Situation

CAWC built their own website, where they now display all their paintings. For each painting they show artist name, painting name, date, size and price. The website includes an order form allowing people worldwide to buy a painting online. When an order is received, the customer is sent a PayPal invoice. Once payment is received, the painting is shipped.

The website also has a form allowing people to sign up for a quarterly newsletter that has information about new paintings and coming events. CAWC sends the newsletter by e-mail.

The collective has started using Artificial Intelligence (AI), building on the techniques of Old Masters, Van Gogh, Munch and Bacon to create paintings. They're aware that, in October 2018, an AI-generated painting sold at Christie's for $432,500.

8.3 Digital Improvements

CAWC claim that, to date, they've used technology to innovate, remove intermediaries, get closer to customers, provide instant information to customers, and allow customers to purchase instantly.

© The Editor(s) (if applicable) and The Author(s), under exclusive license to Springer Nature Switzerland AG 2020
J. Stark, *Digital Transformation of Industry*, Decision Engineering,
https://doi.org/10.1007/978-3-030-41001-8_8

CAWC say that the art marketplace has been democratised. All artists can now offer their works online. Art buyers no longer have to go through art dealers and auction houses. They can buy affordable paintings and objects online.

CAWC say that digital technologies such as computer graphics, Virtual Reality and 3D printing have revolutionised traditional views of art, enabling artists to expand the bounds of creation and develop new experiences such as total immersion in a painting. The Collective wants to participate in collaborative studies and other activities that address the emerging world of digital art. There are many issues to resolve in many areas such as creating, handling, displaying, preserving, collecting, provenancing and sustaining digital art. There are also legal issues, contractual issues, and intellectual property issues. Examples include: identification of original digital art, the artist's signature; registration of original digital art in a library; documentation of original digital art; display of original digital art; making and valuing of copies; maintaining legacy digital artwork for which the initial hardware and software systems no longer exist; and security of original digital art. There are many associated questions: Is a digital work of art the underlying program or the corresponding audience experience? If a digital work of art is displayed in a different environment, is it a different work of art? How will we evolve from traditionally passive and common art museum experiences to interactive and personalised art museum experiences? How will haptic experiences be documented? How can Art students best learn in a world of digital skills, digital knowledge, digital processes, digital communities and digital culture? What governance, policies and guidelines will be needed to support digital art activity?

In response to the changes in the art world, CAWC has developed a multi-axis digital transformation strategy to move further forward.

8.4 Moving Ahead

Now CAWC is implementing its multi-axis digital transformation strategy.

The first step on the training axis will see CAWC posting video lessons on YouTube. Then it will offer a free Massive Open Online Course (MOOC), "Art and Sculpture 4.0". Students from around the world will be able to participate in the course. They'll have the option to receive, for $100, a personalised course verification certificate. The best students will then have the option to enrol for an onsite "Art and Sculpture 4.0" course. High achievers may then be invited to join the community.

Another axis of the digital transformation strategy addresses art developed by Generative Adversarial Networks (GAN). The collective sees this technology as a good starting point for AI-enabled art.

A third axis of the digital transformation strategy is development of an AI sculpture robot. This will build on concepts and techniques of Michelangelo, Moore, Giacometti and Tinguely to take sculpture into a new dimension.

A fourth axis of the digital transformation strategy addresses Virtual Reality art. The collective foresees that, in the future, many artists will work in virtual environments, and others will work across multiple environments, both physical and virtual. The use of digital technology within art will greatly enhance creative opportunities. A fifth axis is to use reinforcement learning, a kind of machine learning, in painting. The idea is that the computer will show a long succession of paintings. For each painting, the artist will give a score showing how well or badly the painting meets the artist's envisioned painting. The computer will reinforce what scores high, and play down what scores low. Eventually it will understand the artist's thinking and produce the painting that the artist wants. From there, the artist will be able to computer-generate multiple unique customised versions for collectors.

Chapter 9
DT at a Passenger Transport Company

9.1 Introduction

Springfield Coach and Bus Mobility Company (SCBMC) is Springfield's largest coach and bus operator. It's been a part of Springfield life for many decades. It's been there for several generations of Springfield residents.

9.2 Current Situation

Under the leadership of a dynamic CIO, SCBMC built its own website and apps. It also communicates on Facebook and Twitter.

The SCMBC website has a mass of information. It includes:

- network status;
- service updates;
- information about each coach and bus line;
- information about each coach and bus stop;
- information about each coach and bus station;
- information about all ticket vending machines;
- 2D and 3D Maps showing all lines, stops and vending machines;
- videos of all lines, stops and vending machines;
- fares, standard and special tickets, and travel cards;
- offers and concessions;
- tickets for groups, special offers for seniors, families and tourists;
- the Frequent Traveller program;

© The Editor(s) (if applicable) and The Author(s), under exclusive
license to Springer Nature Switzerland AG 2020
J. Stark, *Digital Transformation of Industry*, Decision Engineering,
https://doi.org/10.1007/978-3-030-41001-8_9

- purchase and payment options;
- timetables;
- night-time services;
- a real-time trip planner;
- special journeys;
- park and ride;
- a price list of all products and services;
- information about accessibility;
- information about environmental footprints;
- infrastructure repair schedules;
- press releases;
- news;
- rules and regulations;
- contact information;
- a query form;
- a complaint form;
- an online purchase form;
- Frequently Asked Questions;
- About Us;
- a page titled Join Our Team.

At each coach and bus stop, there's a display with up-to-the-minute information (e.g., time, availability of seats) on the next vehicles arriving.

In each vehicle, there's a "black box" that collects sensor data and monitors location and performance. There are sensors reading air and road surface conditions. There's up-to-date information about the vehicle's progress (e.g., time of arrival at next stops, availability of connections). There's video recording of the inside of each vehicle, and video recording of the road ahead of and behind each vehicle. There's an automatic ticket scanner for fast entry. The number of passengers boarding and deboarding at each stop is recorded. Each vehicle is equipped with solar panels, wi-fi, and sockets at each seat.

Passengers can access information from their smartphones. They can see when any vehicle will arrive at any stop. They can see the status of their connections.

Drivers are identified by fingerprint/retina scanners at the start of each journey. They take a breathalyser test before starting the vehicle. They can see their work schedule on their smartphone. They can input data about traffic conditions. They get instant information about the miles they've driven, and their bonus payments for special journeys.

Each coach and bus depot maintains information about the vehicles in the depot, and information about vehicles being maintained. Each coach and bus depot is equipped for Web surveillance.

The traffic control centre can instantly display the location and performance of each vehicle. It analyses "Big Data" from vehicle black boxes to optimise performance. It tracks the location and miles travelled by each driver, making sure they don't drive further than allowed in each work period.

9.3 Current Projects

Discussions are taking place with insurance companies to benefit from lower insurance rates as "black box" information is available.

Discussions are underway with vehicle manufactures for SCBMC to be able to lease coaches and buses on a "per mile" basis, instead of purchasing them outright.

9.4 Current Issues

SCBMC is struggling to keep up with demand. Vehicles and infrastructure are operating at the limits of capacity. However customer expectations are increasing. Travellers want to prepare their travel in advance on their smartphone. They want to be able to make a single booking for tickets for their entire journey regardless of which transport companies are operating each of the legs of their journey. Travellers want on-time co-ordinated travel across the multi-operator network. They expect real-time information. They want immediate notification of delays, and reliable advice on available options.

There's a whole host of customer complaints. Customers say they're confused as the website doesn't have the same look and feel as the apps. Customers are annoyed because they have to re-enter the same data in different apps. (That's because the apps were developed separately.) Customers say that if they ask company staff a question, they're told to look on their smart phones for the answer. Staff don't have deeper knowledge than the apps. Customers say that invariably their favourite functions are removed whenever there's a new version of an app. Customers say that some of the information on the apps isn't up-to-date. For example, a bus stop will be moved, but the app will still show its previous location.

Customers say that frequently a vehicle will arrive late, and then they have to re-plan their travel. Sometimes that means switching to another transport company. But the app doesn't adjust their ticket price for any ticketing differences. It makes them cancel their old reservation, and then make a new reservation. But the app doesn't take account of the price reductions they had for the old reservation, and makes them pay full price for the new reservation.

9.5 Looking Forward

SCBMC has been surprised by the rapid change in customer expectations in the last few years. The company's leaders put it down to the swift spread of smart phone apps into everyday life.

SCBMC hadn't foreseen this new digital world. In previous years, they hadn't thought about what the future might look like. The change seemed to have come from nowhere. Now the new world was here, but SCBMC had no plans to address it.

SCBMC's leaders decided they needed to make a new start. The first step had been to fire the CIO. This was a difficult decision as executives had placed unprecedented faith in the CIO.

They had been delighted by the way everyone had responded to the CIO's appointment and approach. The website, which was widely seen as one of the best in the industry, had won numerous awards.

However, it seemed as if the CIO hadn't been able to build on that success and take SCBMC further on its digital journey. The CIO clearly hadn't foreseen the rising impact of digital technologies. This had led to customer hostility. And within the company there was a mutinous mood in the IS Department, with members frustrated by the CIO's failure to address such a key issue. People said the CIO had been freeloading and had become too involved in self-promotion and self-enrichment at conferences sponsored by technology vendors. Many ideas were just ignored, such as: using AI to improve vehicle safety in traffic; offering passengers the option of e-tickets; letting passengers pay with NFC (near field communication) devices; giving passengers information about routes which were crowded and should be avoided; using AI to simulate proposed changes to routes, timetables and other operations without disturbing passengers; allowing passengers to pay with apps like Google Pay and Apple Pay; using Bluetooth Low Energy (BLE) beacons to guide visually impaired passengers; using high-density screens to display richer content and improve the passenger experience; going multi-modal by integrating e-bike docking stations, taxi stands and local trains into the network; giving passengers information about street pollution levels so they can avoid highly polluted streets; selling passenger footfall Big Data to shops and restaurants so they can optimise their product and service offers and staffing; using facial recognition technologies to identify, fine and prosecute non-compliant passengers; playing Happy Birthday on the vehicle speakers when a passenger celebrating their birthday boards; giving passengers a gift after they've made a certain number of journeys and travelled a certain distance; improving driver efficiency with feedback on energy usage, emissions and punctuality; and increasing passenger security with video recording of the area around each coach and bus stop.

A search was on for the CIO's replacement. One of the first tasks for the new hire would be the development of a Digital Transformation strategy, and the rapid implementation of the corresponding solutions.

Chapter 10
Digital Transformation at a Hospital

10.1 Introduction

Springfield District Community Hospital (SDCH) is a regional hospital with over 500,000 patients on its books.

10.2 SDCH's IS Systems

SDCH has invested heavily in IS over the years. It has IS systems for administration of the hospital, and for staff and patients. There are systems for financial and personnel management, material management, clinical management, and pharmacy management. There are specialised systems such as the radiology system and the laboratory information system. There's a patient portal which enables on-line appointment scheduling, check-in, discharge, post-discharge care and home care. And then there's the electronic medical records (EMR) system.

And then there's the hospital's web site. This has a mass of information for patients, for visitors, for health professionals and for hospital staff. In total, the web site has over 700 publicly accessible pages. For patients, there's information on nearly a hundred services that the hospital provides. There's information on the location of each service, along with opening hours, and contact numbers. There's information about medical conditions and frequent treatment procedures. There's information about being an in-patient and an outpatient. There's information about consultations, and about changing or cancelling an appointment. There's information about appointment reminders, which can be sent either as a text message to a smartphone or as a voice message to a home phone. The site also has a section focused on patient experience, enabling patients to provide feedback such as complaints, comments, compliments and improvement suggestions.

J. Stark, *Digital Transformation of Industry*, Decision Engineering, https://doi.org/10.1007/978-3-030-41001-8_10

For visitors, there's information about hospital buildings, wards and their visiting hours, as well as about useful things to bring, and an outline of staff roles and responsibilities. For health professionals, there's a directory of the hospital's consultants, and its nationally-rated areas of clinical excellence. There's password-protected access to information systems about patients, often limited to particular organisations and only accessible from secure networks. For hospital staff there's a separate password-protected zone of the website. The staff zone includes information such as staff calendars, checklists, awards, learning opportunities, general health and wellbeing, and ways of dealing with stress.

The site has a section about the hospital, its history, organisation and people. The site shows how the hospital is run, its vision of "great healthcare for you from our people". There's also a section about research and innovation at the hospital, and its role in the education and training of doctors, nurses and other healthcare professionals. The site shows several ways for people to get involved with the hospital, for example for work experience, or as a volunteer, or as a fund raiser, or as a Friend of the hospital.

The site also has a News section, and a Latest Vacancies section. The hospital is present on social media with Facebook and Twitter accounts.

10.3   Digital Transformation

Last year, the hospital's CIO was visited by a sales rep from one of their IT providers and told about digital transformation. It sounded interesting, so the CIO set up a Digital Transformation Team, including representatives of the IT provider, to explore how to take advantage of digital transformation features such as mobility, cloud computing, big data analytics and artificial intelligence.

The Digital Transformation Team identified several issues. For example, they found a siloed approach between each department, such as Ambulatory, Critical Care, Emergency, Obstetrical, Perioperative and Psychiatry. They identified over 200 IS applications were used by the hospital, although some hadn't actually been used for years. They found that many people hadn't been trained to use systems. They found some people had more than ten different logins for different systems. They said that communication was slow, bureaucratic and inflexible.

The Digital Transformation Team reported patient discontent with some patients saying they felt like a number, not a person. And that they felt they were seen as necessary evils to support the hospital's internal organisation. Some patients complained that they had to repeat multiple times, to different staff members, their symptoms, feelings and medical history.

10.4   Digital Transformation Opportunities

The Digital Transformation Team identified many new technologies and many opportunities. They included them in what they called the Pole Star Proposal report, as it had ten action lines, and showed the way forward. The technologies identified included machine learning, robotics, mobility, wearables, telemedicine, predictive analytics, smartphones, the Internet of Things, Cloud, Big Data, Analytics, and Artificial Intelligence.

The Digital Transformation Team proposed several improvement areas including a Patient Lifecycle Information System (PLIS) to improve the doctor-patient relationship. This system would make it possible to track a patient throughout their lifecycle, including both as an in-patient and as an outpatient. It would track them from cradle to grave, from check-in to check-out. The system would better inform each patient about their treatment and care by providing real-time digital information at the bedside. The system would be used to schedule appointments much more efficiently, and get patients to come to follow-on appointments. It would also remind patients to take their medication, and inform them about healthy behaviour and lifestyle. It would enable doctors to write prescriptions from an app on their smartphones, and improve administrative functions with self-reporting dashboards. The system would also enable doctors to take notes, photos, and videos from their smart phones. Enhancing communication and collaboration, these could be shared, seamlessly and securely.

The Digital Transformation Team proposed a Consolidated Healthcare Database. This would provide, for each patient, full information on patient lifestyle, habits, families, social environments, diseases and other complaints, allergies, medication, immunisation status, radiology images, pathology and cardiology reports, laboratory test results, sex, age, height and weight. A Digital Twin of each patient would be automatically created from the database. The team also proposed stripping paper out of the SDCH organisation to make it more efficient. A scanning solution would automatically digitise all existing paper-based records.

The Digital Transformation Team proposed a Digital Decision Making System. Digital data would be analysed in digital centres to enable decision-making and discover trends, patterns, and correlations. Predictive analytics would be used to analyse health records and patient data, identify patient risks for numerous conditions, and give real-time insights to support immediate and accessible care.

The Digital Transformation Team proposed a Wearables Strategy enabling continuous clinical monitoring. Patients would wear fitness, glucose, sleep, exercise, temperature, weight and blood pressure monitors to manage their health. Readings from these monitors would automatically update the patient's Digital Twin. The wearables would also enable nurse calls, panic alarms, notification of care providers and real-time status reports.

The Digital Transformation Team proposed an Industry 4.0 Strategy. Robot Nurse Assistants would allow nurses to spend more time providing care. Robot Porters would transport linen and food, allowing patient escorts to spend more time on patient transport. 3D printing machines would be used to make customised replacement parts for patients. And patients would be able to interact with a Virtual Doctor. Robot surgeons would carry out remote surgery.

The Digital Transformation Team proposed use of Artificial Intelligence techniques in areas such as diagnosis, treatment recommendations, and decision-making for prescriptions.

The Digital Transformation Team proposed an E-medicine Strategy. Engagement with patients on social media would be increased. Online chat and Digital Assistants would be provided for instant advice. E-consultations and E-check-ups would be available for outpatients. Secure e-mail communication with doctors would be enabled. There would be interfaces to Connected Homes. Texting would be used for reminding those required to take medication at home. Mobile devices would alert care providers to emergencies at home, and enable requests for emergency assistance.

10.5 Medical Staff View

The CIO hadn't included anyone outside the IT Department in the Digital Transformation Team. When other people in the hospital, such as surgeons, doctors, nurses, technicians, lab staff, pharmacists, psychiatrists, dieticians, social workers and patient transporters heard about the proposal they protested.

They claimed that the transformation should be of the existing resources and their organisation. They said these are out-dated, and better suited to the 19th century than the 21st century. They said that computerising the hospital's current activities would bring no benefit to patients. They said the hospital's mission was to improve the health of the population, not to administer and computerise whatever was in sight. They wanted the transformation to reflect the many changes in both the population and treatments.

Their objectives included moving from a healthcare model of treatment to one of prevention, faster and more effective treatment in hospital, shorter hospital stays for patients, and improved home care enabling seniors to stay in their homes as long as possible.

These objectives led them to a vision of a well-equipped central medical unit, several smaller satellite medical units and mobile health units working in the community. For the central and satellite medical units, they saw a need for well-planned installation of equipment such as Magnetic Resonance Imaging (MRI) scanners; Computer Axial Tomography (CAT) scan machines; 3-D mammography machines; particle accelerators for treating cancer; portable X-ray units; robot surgical machines; life support equipment; monitors such as for electrocardiography and electroencephalography, and laboratory equipment for analysis and diagnosis. They proposed investing in such health improvement equipment, some of which cost millions of dollars, rather than financing huge IT projects of no clinical benefit and likely to fail. They also pointed to the cost of procedures such as heart-lung transplants; bone marrow transplants; open heart surgery; tracheostomy; and gene therapies.

Medical staff say that key performance indicators (KPIs) for their vision, such as length of stay in the hospital, treatment success rate, readmission rate, total days in hospital, significant incidents, patient satisfaction and staff turnover, would be easy to identify and measure. And any transformational approach should support that vision, with any proposed IT solutions being linked directly to these KPIs.

They also pointed out that Digital Transformation initiatives need much more than IT hardware and software. They said that the majority of the investment needs to be in changing the roles, activities and mindsets of people, and making structural changes to the organisation. They say that not only do IT investments nearly always turn out much higher than planned, but related training costs and maintenance costs will also be high.

10.6   Resolution

The difference of opinion between IT staff and medical staff has not yet been resolved by hospital administrators. They are currently not sure what to decide and, looking for the most valuable outcome, are weighing up the costs and benefits of different approaches. Hospital administrators asked an independent expert to review the Pole Star Proposal and inform them of potential risks. The following risks were identified by the expert.

Risk 1:   Unachievable Hospital Management System. There was a high risk that the proposed high-tech hospital management platform was too complex to build and would never be fully implemented. There was also a high risk that it would greatly exceed its budgeted cost.

Risk 2:   Unmanageable Huge Data Volume. The volume of data to be managed in the future under the Pole Star Proposal was gargantuan, tens of times more than at present. The Pole Star Proposal did not address its management. There was a high risk it would be uncontrollable, leading to disastrous results and effects.

Risk 3:   Defenceless against Cyber Attack. With the hospital being a very open environment, there was a high risk of being unable to protect systems against cyberattack. The Pole Star Proposal did not address cybersecurity.

Risk 4:   Poor Medical Device Interfaces. With a plethora of new medical devices being rushed to market with little thought given to user experience, there was a high risk of health professionals being unable to use the hospital's future equipment correctly.

Risk 5:   Patient Data Confidentiality. With the hospital being a very open environment, there was a high risk of being unable to maintain the confidentiality of patient data. The Pole Star Proposal did not address confidentiality of patient data.

Risk 6:   Part-time Workforce. In today's gig economy, with short-term contracts replacing permanent jobs, there was a high risk that many temporary health staff would not be able to use the hospital's future systems correctly.

Chapter 11
Digital Transformation at a Bank

11.1 Introduction

Springfield County Regional Bank (SCRB) is a regional bank with millions of customers. SCRB works with its customers to achieve their financial goals. Through superior performance it aims to create exceptional value for its customers, staff and investors.

11.2 History

SCRB is on its third attempt at Digital Transformation. We hope we've got it right this time.

We first looked at Digital Transformation when we heard from industry experts that banking was changing dramatically with disruption caused by new technologies and the digitisation of the world. This fitted with what we were seeing. We saw declining branch traffic. We saw competition from newcomers, fintechs, not our traditional competitors. We didn't know which products and services to develop, which to ditch. But, not being a global mega-bank we weren't able to throw hundreds of millions of dollars at digital transformation.

For our first attempt at Digital Transformation we decided to move all our systems to the Cloud. At the time, that's what everybody was talking about. It seemed the obvious and easy thing to do. After a few months we stopped the project. It was going nowhere. We realised that we didn't really know what we were aiming for, or why.

For the second attempt we decided to do it differently. We picked two of our best upcoming high-potential VPs. They both had more than twenty years with the bank, great credentials, deep understanding of banking rules and regulations and of our way of working. They were just as good at new high-value customer acquisition and maintaining existing client relationships as they were at cost-cutting. We gave them free rein, one in our branch business, the other to start a new digital business competing with one of those new fintech companies. But within a year, both of those endeavours had stopped. One of the execs quit. Fortunately the other stayed on, and she helped us a lot to understand how we should move forward. We realised that we needed to think differently. Some of us had to disengage from everyday running of the business, go back to basics, and really understand our business.

11.3   Background

We're a big regional bank with millions of customers. Until recently, we've had a loyal customer base, community support, a range of great products and services, and a reputation for security. We've grown by acquisition over many decades, which is why we now have so many different legacy computer systems. That's what caused the trouble when we tried to move to the Cloud. It turned out we had to unravel something like a giant bowl of tangled spaghetti, and we never made it. We didn't have the resources to do it.

Another issue we have is that we have several different channels to our customers. There's the traditional branch business. Then we introduced the Automated Teller Machine (ATM) channel. The next channel came when we introduced e-banking, which is more or less its own world. And most recently came the mobile business channel with the apps. These days, all four run in parallel. Keeping a common look-and-feel is challenging.

Another issue is that our customers don't all behave the same way, they have all sorts of different profiles. We can classify our customers in many groups according to criteria such as age, wealth and behaviour. For example, by age there are the post-millennial Generation Zs, the millennials, the Gen-Xs, the boomers and the seniors.

Another issue is that internally we have our own special culture and mindset. On one hand, we're very cautious, follow all the procedures, and get our figures right to the exact penny. On the other hand, we sometimes get involved with billion dollar transactions, and behave totally differently, as in the subprime mortgage crisis in 2007 which drove the world into recession.

Putting all this together makes for a complex environment, to which several layers of regulation are added. However, one thing is clear, we don't want to lose any of our millions of customers, and we do want to add many million more. So when we started our third attempt at digital transformation we decided to take a close look at our customers and the services we offer them.

11.4 Services

We don't offer customers all that many services, mainly account management, deposits, loans, payments, mortgages, overdrafts, credit cards, and guarantees. When we started looking in detail at some of these, we were surprised by what we saw. None of these services are exactly rocket science, but we saw that often we'd ask the customer to fill in all sorts of forms and provide secondary evidence in the form of paper documents, and then weeks would pass before we got back to them.

Our VP who we'd asked to start a new digital business was working on this. After a while she started saying there was nothing much wrong with our legacy systems, they could do whatever we asked in a few milliseconds, the real problem was our internal workings. That didn't go down well with a lot of senior people who'd been responsible for setting up our internal procedures. They started justifying why things had to be done that way. And they'd say that you can't create a customer relationship with a smart phone. That played into the hand of our VP, as she'd been analysing all the data the regulators and government makes us keep, and would come back with an answer like "you're so right, but according to our recent Big Data analytics report, 98% of customers using our apps have average account balances of less than $500, and on average last visited a branch eight months ago, whereas there's an executive assigned to all customers in the wealthy segment, and on average they've met them in the last three weeks."

11.5 Customers

So then we really started looking at our customers in detail, looking at how the services we offered them actually worked in detail, all this based on the data we had about them. And we do have a lot of information about them, like name, address, date of birth, family members, lifestyle preferences, phone numbers, account balances, outstanding loans, financing deals, mortgages, employment, salary, investments, average monthly spend, home and car ownership, home improvement spend, health care plans, life insurance spend, cash withdrawal profile, repayment history, and total wealth. We were shocked when we looked at the average time and cost of some of our service operations. They may have been normal, industry-typical twenty years ago, but they were way behind today's norms in the digital world.

For example, it was taking days to open an additional account for an existing customer. But when we looked closely we saw it required at most a few seconds of processing time. The rest of the time went on looking for documents and waiting for approval signatures. And then we thought, is this necessary for an existing customer? Why not just check a couple of key figures? And then we saw that if the customer is online we can send a confirmation mail with the new account details. The customer will be impressed, and won't even think of checking out the competition. Then we realised that would also work if the customer was in a branch or mobile. It's a transaction that just takes a couple of seconds. As someone said, that means we can respond a hundred thousand times faster than we used to.

Somehow something clicked, and you could almost see people changing the way they were thinking. If we can open an additional account in a few seconds for an existing customer, how about if they ask for a loan? So we looked in detail, and saw that we had so much information on most customers that we would be able to automatically run through most of the loan processing steps. Then we started looking at the other steps in detail, to see what was missing, and what we'd have to do to get the app to execute all the steps without human intervention. We'd find different answers for different steps. Sometimes we actually had the missing information available digitally in the bank, but on a different computer. Sometimes we had it on paper in a vault somewhere, so we realised we'd need to digitise some documents. Other times, we could get the information from someone else's database, maybe a government database.

Then we looked at mortgage processing. We quickly worked out how we could respond twenty times faster, yet offer a personalised mortgage with financing options to fit the exact customer need.

11.6   A Changing Environment

You could feel some people were getting nervous, wondering if their job was going to be digitally transformed away. That hasn't happened so far.

Yes, there may be a need for fewer people in some areas, but at the same time there's a whole lot of work to do in areas such as implementing the changes, digitising and cleaning up existing documents, training and retraining, fraud detection and management, default management, risk management, customer segmentation, consumer engagement, customer retention mechanisms, and improving reporting to customers.

There's also cross-selling, relationship building, knowledge management, digital marketing, improved communication, improving our social media presence, managing trust and reputation, making explanatory videos of our services, addressing data privacy, confidentiality and security issues, executing face-to-face activities better, and interacting more effectively with customers by phone. With many new activities, and many activities to improve, we realised that we had to invest in training and retraining for our people. But that investment didn't stop when we thought it would. The pace of change was so high that we started providing regular ongoing support and learning programs for our people. That got us thinking about ongoing support and learning programs for our customers. They're also faced with digital change. They're moving to e-banking, mobile banking and Finance 4.0. When they make the change, it's a new world for them. Some of our people thought customer training wasn't necessary as our chatbots are in constant use. Others said that chatbots are great at what they do, but there's a lot they don't do. For example, they miss all the visual signals and body language you get when you're in front of a customer. We did some tests with a selection of customers. Based on the results, we decided to roll out explanatory videos and training sessions to help customers. We've had good participation and feedback from customers. They like these sessions while, from our side, we see training sessions as a great way of getting to meet customers and help them. These sessions are also a great way to find out what customers want. Discussions during training sessions led to our electronic Personal Financial Assistant which helps customers with taxes, secure online shopping, remembering passwords, and making foreign currency payments during vacations. We also found out from training sessions that the AI unsupervised anomaly detection approach we'd implemented to detect fraud didn't work the way it should have done. So we withdrew it, and will be more careful with machine learning technology in future. Customers come first, technology comes last!

11.7 Technology

You could feel some people in the bank wondering about the information technology they'd heard about. Technologies such as Artificial Intelligence, Machine Learning, Cloud, blockchain, Internet of Things, chatbots for interacting with customers, and smartphones for branch counter staff. Shouldn't we be acquiring and using these technologies?

We're well aware of these technologies, and keep an eye on their development, but we've learned that most important is first to understand in depth how we work with customers and then to identify how we could work better with them. And only then to look to see which technologies could help us.

11.8 Going Forward

Going forward, we're looking more at new business opportunities, customers, products, services and geographies, than at new technologies. We know our customers well, so we should be able to offer them personal recommendations, give them more choice, propose a range of investment products and services with revenue and price comparisons.

Knowing our customers well, we'd like to cross-sell them other financial products. We won't even need to develop these products, for example we're working closely with several fintech partners so we can relabel and tailor their products and services and offer them to our customers.

Going forward, the first priority is to generate more revenue from existing customers by identifying and offering the value-adding products and services they want.

Going forward, and this is very much a second priority, we also want to generate revenue from new customers looking for a bank that's a leading player in digital banking, offering value, quality, security, convenience and a great experience. We know there are millions of people out there looking for a new bank they can trust. But that's not currently our main thrust because, at this time, we don't have the systems in place to provide the service they'd expect.

Chapter 12
Digital Transformation of a Retail Store

12.1 Background

Springfield Community Home Stores (SCHS) is a private family-owned business. It's a non-profit corporation, and its articles of incorporation lay down that no dividends will be paid, and no debt taken on. The founders knew that would limit expansion, but they wanted to focus on serving the local community. The company now has five stores that together serve more than two million people. These people, living in a mixture of inner city, suburban, small town and rural areas, are less than 30 min from a SCHS store.

12.2 Not a Leader

SCHS doesn't claim to be a retail leader. It watches the leaders from afar, identifies what works, and then follows a year or two later. That way it can see what works, and what doesn't. That way it can reduce risks significantly, yet is never far behind.

So, for example, SCHS has a Customer Relationship Management system that's reasonably good. There's an SCHS credit card, and a loyalty program. Customers can use their store card to gain points, use electronic coupons and benefit from special discounts. SCHS has a good stock control system, so when SCHS introduced an online store, it was able to show online customers which articles were in each store, and how many were available. SCHS now has 15,000 products online.

SCHS offers Click & Collect, so customers can order goods from the online store, then pick them up from the nearest bricks and mortar store. Alternatively they can be delivered to the customer's home. In the stores, SCHS has introduced self-serve checkouts, and enabled contactless payment, digital transfer, and digital wallets. In 2017, SCHS launched mobile applications on various platforms. That meant customers were offered the three channels of web, app, and bricks and mortar.

On social, SCHS has Instagram, Facebook, Twitter and Pinterest accounts.

12.3 Many New Technologies

SCHS is seeing a lot of new technology solutions being proposed. There are new technologies such as Artificial Technology, Chatbots, cognitive learning, advanced analytics, blockchain, mixed reality, immersive, radio-frequency identification (RFID), Internet of Things (IoT), Near Field Communication (NFC), and Bluetooth Low Energy (BLE).

SCHS has seen many exciting new solutions being proposed. These include: scanners that ascertain a customer's face contour and skin colour, then use AI to suggest hairstyle options and make-up colours; customer avatars that try on clothes and, in seconds, show how the customer would look; use of augmented reality technology to show how customised product features would meet the customer's unique requirements; in-store dining experiences, with beacon order-taking and table-selection, and robot service; digital price tags; interactive shop window displays; as well as automated shopping carts that follow the customer, automatically avoiding collisions and showing the check-out total. Among other proposed solutions are:

- digital kiosks in the bricks and mortar stores to enable product search by customers;
- service robots as sales assistants;
- virtual changing rooms in which customers can virtually try clothes on in an online store;
- in-store Wi-Fi access to improve customer experience;
- use of cryptocurrencies such as Bitcoin;
- drone-based deliveries;
- RFID chips to make the supply chain work better;
- sales associates equipped with tablets that show key data about the customer they're serving, and propose special financing deals;
- smart mirrors;
- virtual catwalks;
- digital runway shows;
- in-store personalised promotions and recommendations sent directly to customers' smart phones;
- automatically discounted prices for loyal customers;
- improved collaboration with suppliers across the digital supply chain;
- more productive Digital Marketing campaigns promoting products and services through digital media;
- advanced analytics suggesting a personalised experience for each customer;
- chatbots answering customer's online queries;
- in-store Virtual Assistants helping the customers;

- voice-activated technologies to enable selection and purchase of products in the online store;
- analytics that take account of customer lifestyles;
- provision of pre- and post-purchase support across the purchase lifecycle;
- Advanced Virtual Reality giving customers a fully immersive online shopping experience;
- Beacon Technology that identifies customers when they enter the store and then pushes customised notifications to their smartphones.

12.4 Many Questions Raised

SCHS is excited by the new technologies. It sees their value and importance. However, it's not sure what's most important for the company. It's asking a few questions. How would these technologies affect us? How much would they increase revenues, how many millions would they cost? How long would they take to implement? What should we start with?

12.5 CEO Input

The CEO of SCHS said that before investing in new technologies, they needed a Digital Transformation concept. He felt the company lacked knowledge and understanding of Digital Transformation. He said he didn't want inflexible unconnected technology, and didn't want different parts of the business working at cross-purposes.

12.6 CEO Decision

The CEO decided it would be best to start by asking customers what they liked about the company, and what they would like as improvements. Some of the customer feedback was instructive.

"Before leaving home, I look at your webcam to see if there'll be room for me to park. I feel safe with the security cameras."

"I love the home delivery service. When I know I'll be at home, I take the free option. Sometimes, when I've been very busy, I've paid extra for fast delivery, but it was worth it."

"I really appreciate the help your sales folk give. They're so friendly and they give such good advice."

"Shopping at SCHS is a great way to meet new people."

"I usually shop with a friend, and we like the coffee bar. My friend likes the Caramel Macchiato, I adore the Café Mocha"

"I like your returns policy. It's not often that I return something, but it's good to know it's there."

12.7 Survey Response

One of the questions asked how the customer would react if SCHS closed the bricks-and mortar stores and just had an e-store. Surprisingly few people wanted that. One customer replied, "Please keep your current stores with real salespeople. I've bought books online, but even then I miss the assistance of a knowledgeable sales associate. They can be so helpful. And in your stores, it's so easy to look round and see useful things to take home." Another customer said, "I just love the simple, easy shopping in your stores. I don't want to inconvenience myself downloading dozens of apps. I don't want to feel bad about myself not remembering how they work." Someone else responded, "The only thing I don't like about your stores is that sometimes it takes so long to pay. Couldn't I just pay from my smartphone without queuing up?" Among other responses were:

"Some things I buy on line, standard stuff like soap, but not really individual things like clothes. I want to touch the clothes I'm buying, and see the colour. I want to feel the material, feel its thickness. I want to try it on to feel if it's a great fit, and see how it moves."

"I've bought food online sometimes, and that's been OK, but I prefer being here and looking round to see what's best. I drop in after work. It's on my way home."

12.8 Strengths

The survey may not have shown a clear path forward to Digital Transformation, but it did highlight SCHS's existing strengths of a loyal customer base, convenient locations, knowledgeable store associates, an appreciated delivery service, and everything under one roof.

SCHS saw these as areas to investigate in more detail for Digital Transformation as they could help it maintain and improve its position against other retailers. It also saw the need for better customer segmentation, better understanding of customers and improved communication with customers. It's now thinking about taking a cautious step-by-step approach to Digital Transformation in stores. It may start by introducing new digital concepts in just one of its stores to understand how customers respond and interact with specific technologies, services and layout. This way, if something works well, it can then be rolled out in other stores. If it doesn't work, it can be quietly removed. The company is thinking about taking a similar step-by-step approach to its online services for customers. Starting with some important basics such as hiring more people with digital skills, improving quality assurance testing, eliminating website downtime, ensuring pages load fast, improving cybersecurity, optimising website journeys, expanding customer card features, and broadening social functionality. Meanwhile, in parallel, the company will be working on developing its Digital Transformation concept and Digital Transformation strategy.

Chapter 13
Digital Transformation of Springfield Police Force

13.1 Background

Springfield Police Force's Commissioner always says they started their Digital Transformation journey long ago. They're now far from the days when officers arrived on the scene long after a crime was committed and wrote notes on paper with pencils, then went back to the station house to type up paper reports they put in paper folders. Making using of the huge volumes of digital data that are now available, the Commissioner says policing in Springfield is changing from being reactive to being proactive. It's becoming easier to identify where crimes are likely to happen and then prevent them.

Reflecting changes in society, the scope of police activities has also changed greatly since the old days. Now it also includes terrorism, cybercrime and an online presence. But the mission hasn't changed, it's still to keep Springfield safe for everyone. The objectives of the Springfield Police Force (SPF) include: maintaining and enforcing law and order; protecting people and property; and preventing, detecting, investigating and solving crime.

To continue to address these effectively, the Commissioner recently launched a new Initiative, "Safe and Secure Springfield 2025". This takes the following multi-dimensioned approach.

13.2 Self-Service Police Portal

The portal provides the community online information about the SPF and its mission, services, leadership, technology, patrols, policies, bureaus, statistics, careers, academy and community programs. It includes information about law enforcement

J. Stark, *Digital Transformation of Industry*, Decision Engineering,
https://doi.org/10.1007/978-3-030-41001-8_13

topics such as victim rights, property protection, lost and found, and how to locate a towed vehicle. It can be used to report a crime, to pass on a tip, to report a missing person, to pay a fine, to request a person verification, to verify an address, to request permission to hold a public event, and to report a vehicle theft.

13.3   Integrated Regional/National Policing

Police work is all about information, and there's a need to exchange information across force borders and with many different organisations such as hospitals, doctors, the fire department, and transport providers.

That's why the SPF is integrated with local, regional and national central and distributed police data systems and databases. Seamless interfaces and coherent information access improves efficiency and reduces costs.

13.4   High-Tech Connected Frontline Policing

The SPF has a wide range of resources. One thing they all have in common is that they're connected.

Handheld mobile devices provide officers and staff with a multitude of services: accurate information at their fingertips; translation services; ticket-writing; information recording and dissemination; identification of suspect objects and chemicals; and information about surrounding buildings and vehicles. They have apps for facial recognition, fingerprint scans, retina scans, GPS location, identifying the nearest available cell and real-time route planning. Body worn video footage shows exactly what happened in any situation.

There are also apps for use by members of the community requiring assistance, allowing them to call for help. The app automatically provides location information, enabling fast response.

Automatic license plate readers in patrol cars are prevalent. They analyse license plates on every vehicle in sight, reporting stolen vehicles and speed violations.

Police video kiosks are available in many locations to answer community members' questions regarding police services.

Surveillance drones patrol Springfield 24/24 providing real-time information to watching analysts and police dispatchers for fast incident response. Real-time video is sent to crime analysts who send vital information about crimes in progress to officers and warn them of dangerous situations to avoid. Drones can be directed to identify criminals and follow them back to their bases.

Preventing crime requires eyes and ears on the ground, and the SPF doesn't have the resources to position an officer on every street corner. Robot police dogs play an increasing role, reassuring the public as they patrol Springfield night and day, supporting officers, investigating bomb threats, going places too dangerous for officers, as well as chasing and neutralising fleeing criminals.

13.5 Collaborative Community Policing

It wouldn't make sense for the SPF to try to carry out its mission without the support of the community. The public plays an important role in addressing public safety issues. In Springfield, community policing is focused on close collaboration with the community, the individuals and organisations it serves. Components of community policing include neighbourhood sub-stations, neighbourhood patrols, neighbourhood watch, neighbourhood newsletters, social media, and online police portals.

Technology helps two-way communication with the public through means such as online incident reports, reporting of non-emergency crimes, crowd-sourced neighbourhood information, e-mail alerts, texts, and discussion forums.

13.6 Police Real-Time Online Picture (PROP)

The SPF has a huge amount of data from many sources. Some of the data is long-life, some is real-time. The data ranges from maps and descriptions of premises, to crime reports, patrol calls, dispatching messages, and criminal histories. The SPF brings all this together in the PROP system. This provides a comprehensive picture that's easy to understand, allowing both front line and support officers and staff instant access to long-life and real-time information helping them to respond better in all situations. They can focus in on the area that interests them, and drill down for more information as required.

13.7 Digital Case Files

In Springfield, gone are all the paper notebooks, paper reports and paper folders. Now information is all digital, allowing for fast and easy access and communication.

The SPF's digital case files can include texts, photos, video, audio, e-mails, tweets, electronic witness statements, digital evidence on smart phones, body worn video footage, 999 recordings, CCTV surveillance footage, electronic records of evidence, and even wider digital footprints that may be relevant to particular cases.

13.8 Remote Monitoring

Officers can't be everywhere in Springfield all the time. Sensors attached to drones, robots and fixed infrastructure all monitor what's happening in Springfield when officers aren't present. They stream real-time information back to the station, enabling officers and support staff to watch over traffic and property.

13.9 Predictive Resourcing and Scheduling Analytics

In view of all the data available to the SPF, analytics is playing an increasingly important role.

One part of the SPF's resourcing solution schedules well ahead, taking account of coming events in Springfield and the vicinity, front line and support staff availability and requirements, workplace agreements, past events, as well as historical and current trends for property crimes, gun violence, assaults and traffic accidents. It can suggest where officers should be deployed and when. For example, targeting certain likely crime locations at certain times of the day and night. Another part of the resourcing solution analyses and schedules real-time, taking account of factors such as the current weather and traffic situation, incoming 999 s and actual front line and support staff locations and activities.

Doing all that by hand was slow and inefficient, and kept staff from the mission of keeping Springfield safe for everyone. Digital technology improves scheduling significantly.

13.10 Deterrence and Investigation Analytics

The SPF is using investigative analytics in different ways to support investigations. Investigative analytics reads through the huge volumes of digital data available to the SPF. It's used, for example, to find and follow a criminal's social media tracks, to locate clues, to read millions of telephone records to relate criminal activities to locations, and to interpret images of people and locations found on criminals' smart phones.

The SPF is also using analytics in support of preventive policing to enhance the quality of life of Springfield residents. With access to the huge volumes of data that the SPF has, analytics can mean getting a better understanding of a situation. Then for example, the SPF can make better recommendations about bail conditions, repeat offenders, and prisoners on parole. Analytics helps spot patterns and trends, and identify and recognise suspicious behaviour and activities. They predict and forecast where and when potential crimes are likely to occur, allowing the SPF to deploy extra resources to deter occurrence of the crimes.

13.11   The Way Forward

The Commissioner has identified Digital Transformation as a source of new challenges, and recognises that it must also be part of the solution. Digital Transformation is seen as an on-going process in which the first step is to understand the SPF's working processes in detail. After that the next step is to look at ways of doing things better, understand the information needed and used in the processes, and then look at available solutions. The Commissioner says it's important to understand the SPF's current working processes in detail. The intention of achieving this understanding is to be clear about the objectives of the processes, and the activities they support. That lays out a firm base, a starting point, for moving forward. The intention isn't to automate the current tasks of the current activities. Most likely that wouldn't deliver clear benefits. Most likely that would create new problems. Instead, the intention is to transform current activities into new processes that make use of the digital technology that's available. A new process architecture will be built. Then, knowing what needs to be done, the next step will be to identify the information requirements, availability and quality. The information may or may not exist. If it doesn't exist, where will it come from? If it already exists, is it in the right format, is it in an unconnected system? An information architecture will be built. With the process architecture and the information architecture known, the next step will be to look at available systems and functionality. Key requirements here include flexibility, security, confidentiality, availability, connectivity, and mobility. And finally, the Commissioner says that, last but not least, come the people. Policing work isn't done by technology, it's done by police officers. Their work will change, but they'll still be needed. They'll need wholesale education and training to keep abreast of the new technologies. There will be several channels for this, ranging from traditional linear classroom and e-learning courses to short videos on very specific subjects. The Commissioner foresees a plenitude of digital awareness and training activities for frontline police officers and back-office support staff, including: initial digital awareness events; events explaining the new digital environment for policing; events outlining the skills needed to operate in a digital environment, and describing different digital activities and digital footprints; basics of digital policing, showing how digital technology is applied in communication, intelligence and evidence; and training to operate in the digital world, including role play, with sessions on: search, security, and analysis of digital evidence; gathering information and intelligence from the World Wide Web; recovering and analysing digital data; entering accurate, detailed data at the incident scene; and correctly recovering, handling and preserving electronic evidence. In addition, there'll be advanced training in dealing with serious digital crimes; advanced training focused on use of modern technology to investigate old offences; specialist training focused on recovering and analysing digital surveillance video; specialist training focused on countering use of the internet for criminal purposes; specialist training addressing unlawful activities with cryptocurrencies; and specialist training on adoption of algorithmic decision support tools in decision-making processes.

Chapter 14
Digital Transformation of Springfield Regional Airport

14.1 Background

Springfield Regional Airport has always been a leader, a pioneer of new technologies. We were among the first to introduce e-tickets, digital Flight Information Display Systems (FIDS), automated baggage sorting systems and offer a website and e-commerce platform. We were among the first to offer text message alerts, mobile boarding passes and digital advertising screens.

We have a mass of information on our website, information such as arriving and departing flights, destinations, airlines serving the airport, airport maps, journey preparation, baggage rules, airport access, transport to and from the airport, buses, coaches, trains, taxis, car parks, car rental, valet services, baggage and check-in, lounges, VIP services, priority lanes, security checks, reduced mobility services, shops, restaurants, bars, aviation services (charges, fees, operating rules), non-aviation services (advertising, taxis), marketing, sponsorship, cargo (infrastructure, freight halls, fees, rules), airport strategy, governance, sustainable development, news, job offers, contacts, our apps.

14.2 Reasons for Our Digital Transformation Program

We're currently responsible each year for hundreds of thousands of flights, tens of millions of passengers, and more than a hundred thousand tons of freight. Over the next fifteen years those numbers are expected to double. At the same time, our stakeholders are expecting us to reduce costs, improve efficiency and improve the customer experience.

We're looking to Digital Transformation to help us meet the following goals: increase capacity; enhance safety; increase operational efficiency; reduce our environmental footprint; reduce costs; and meet customer expectations.

J. Stark, *Digital Transformation of Industry*, Decision Engineering, https://doi.org/10.1007/978-3-030-41001-8_14

14.3 Our Airport Processes

Our airport is often compared to a small town, there's so much going on. We have more than ten thousand employees. On a typical day, we look after tens of thousands of passengers. At any one time, we have hundreds of activities going on both landside and airside.

For example, there's the passenger departure process. That includes reservation, ticketing, getting to the airport, getting to check-in, waiting, checking-in, baggage hand-over, shopping, eating and drinking, security checks, passport control, restrooms, lounges, duty free, gate holding and boarding. When you have tens of thousands of passengers going through that process during an 18-h day you have to have it defined in detail—the tasks, the information that's needed, the participants in the process, their roles and so on.

Then there's the passenger arrival process with de-boarding, transfer, passenger flow, immigration control, arrival lounge, retail, baggage delivery/collection, customs, lost and found services, transfer from the airport.

In parallel with passenger departure and arrival, aircraft have to be managed. That could include marshalling, towing, parking, cleaning, crew boarding, refuelling, baggage loading, catering loading, passenger boarding, load control, de-icing, and pushback. And then there's the baggage handling process with drop-off, x-raying, sniffing, sorting, transportation and loading. Cargo handling has a different process starting with unloading, customs control, x-raying, and intermediate storage.

As well as passengers and aircraft, we have a huge amount of facilities and equipment to manage. These include air conditioning units, aircraft starter units, baggage belt loaders, baggage carousels, baggage drop-off stations, baggage sorters, baggage scanners, catering vehicles, check-in desks, communication networks, container loaders, de-icers, departure control gates, drainage ponds, escalators, FIDS, firefighting vehicles, ground power units, jetways, lavatory service vehicles, lounge facilities, passenger boarding stairs, passenger buses, passenger flow signage, portable power generators, pushback tractors, refuellers, retail infrastructure, self-service check-in kiosks, transporters, tugs, water pumps, and WiFi networks.

There's also the Air Traffic Control process, controlling aircraft movement in the air and on the ground. And the process for foreign object debris detection and removal on the airfield.

As well as all the frontline operating processes, there are supporting processes such as personnel planning and scheduling. That includes demand planning, shift planning, personnel disposition and task scheduling. We also have Human Resources Management processes for recruitment, payroll and dismissal. And there are processes for planning, preparation and delivery of personnel training.

We have commercial processes covering negotiations and agreements with all our partners including airlines, transport, retail, catering, government, cleaning and weather forecasting services. And there are financial processes such as invoicing, accounts payable, accounts receivable, balance sheet, asset accounting, financial statements, and treasury management.

14.4 New Digital Technologies

We're aware of many new digital technologies that could help us meet the goals set by our stakeholders. They include augmented reality, cloud computing, mobile technology, Big Data, analytics, social media, Internet of Things, connected products, smart machines, robots, augmented reality, Blockchain, machine learning, Artificial Intelligence and autonomous transport systems.

14.5 Digital Transformation

We defined Digital Transformation as the use of these new digital technologies to improve customer experience and streamline operations.

14.6 Improvement Areas

There are many way in which we can use new digital technologies to improve the customer experience and streamline operations. Our strategic focus for Digital Transformation is in four areas. These are Passenger Experience Improvement, Optimised Scheduling, Smart Facility and Equipment Management, and Air Traffic Control.

14.7 Passenger Experience Improvement

We're working on multiple ways to digitally enhance the passenger experience.

Self-service baggage drop, check-in kiosks and touchscreen boarding terminals protect passenger privacy and save valuable passenger time. Passenger identification is enabled and confirmed by biometric cameras and scans of passports and boarding passes. Advanced provision of facial, fingerprint and retina scans speeds the process further. Passengers can self-tag baggage digitally without the intervention of airline employees. They can pre-order food and beverage items from home for collection in airport restaurants and bars. They can order duty-free goods from home and collect their online purchases in airport duty-free stores.

Digital wayfinding signage enables self-navigation through the airport. Alternatively, passengers can facilitate their journey to security lines, restaurants, shops and gates with check-in robots and smart wheelchairs. Airside robots and airport staff equipped with mobile devices can help passengers further in case of any navigational issues.

Real-time baggage tracking with RFID or conventional baggage tags lets passengers see on their smart phone where their bags are at any time. Bag-tracking apps show when a passenger's bags will appear on the conveyer belt. Self-service lost luggage kiosks enable passengers to rapidly retrieve missing bags without waiting for airport staff to become available.

Passengers can use their boarding passes for automated entry at airport and airline lounge doors. They can update status online, and reserve special facilities and services such as showers and massages.

Digital technologies based on biometrics and computed tomography (CT) scanners lead to shorter security lines. Time spent in Immigration lines is reduced as scans of arriving passenger passports, faces and fingerprints are automatically transmitted to counter agents.

14.8 Optimised Scheduling

Modelling and simulation help planning and scheduling. They optimise passenger flows and airport resources at peak times.

Passenger flows are tracked by sensors. Analytics measure and foresee congestion and help relieve the flow at bottlenecks. Personnel shifts are rescheduled to take account of expected and unexpected changes in passenger flows.

Scheduling algorithms help reduce the time an aircraft is on the ground. They evaluate and mitigate the impact of flight delays due to weather, breakdowns, labour disputes and other events. Take-offs and landings are rescheduled taking account of the actual situation. Arrivals and departures are automatically sequenced to optimise runway flows.

14.9 Smart Facility and Equipment Management

Sensors on our items of ground equipment communicate their location and use, monitor performance, and provide real-time performance information. Predictive systems use this information to increase efficiency, highlight areas for improvement, and schedule equipment maintenance to reduce downtime.

Smart facility and equipment management will help us meet the goals we have set for eleven of the seventeen Sustainable Development Goals described in the United Nations 2030 Agenda for Sustainable Development. Our Environmental Management System (EMS) is ISO 14001 certified. We're committed to reducing emissions of pollutants and greenhouse gases by 2% annually. To limit the impact of air traffic on nearby residents, we're reducing the footprint exposed to aircraft noise by 3% annually. We're also reducing water consumption by 3% annually. Our sustainability strategy requires us to limit the energy required for operations, to produce and distribute energy in the most efficient way, and to give priority to energy from renewable energy sources. Our Energy Management System is ISO 50001 certified. We improved energy efficiency 25% over the last decade, and we aim to carry on with that trend over the next decade. We're not far from a 60% share of renewable energy. Smart sensor-based systems will help us continue to reduce emissions, reduce water consumption, reduce energy use and drive down costs.

14.10   Digital Air Traffic Control

Online sensors, cameras and radar systems in and around the airfield improve air traffic control services. Augmented reality and virtual reality devices help pilots during take-off and landing.

Real-time information is input to automated decision support tools to reduce idle runway time, turnaround times and flight delays.

Digital technologies ensure safe and efficient operations on the airfield and in the air.

14.11   Rules for Digital Transformation

Based on its experience, SRA recommends the following golden rules for digital transformation.

Work closely with your airport partners, the airlines, retailers, caterers, ground transport providers, government agencies, and labour unions.

Start by defining and documenting the big picture. Take the time up front to understand and clarify your airport's strategic goals for digital transformation.

Once you have the big picture for your airport, identify a small number of very important digital transformation initiatives. It's not realistic to try to do everything at once. Only select digital transformation initiatives that fit your strategic goals. Don't do stuff that looks cool, but doesn't fit to one of your strategic goals. Build and document the business case for each of your digital initiatives. Don't start with everything at once. Start with projects that fit with your strategic goals and are expected to give results quickly.

Craft a change management plan to guide everyone in your airport through digital transformation. Make sure they all understand the need to transform, and the transformation initiatives. Make sure they have the skills to succeed in the new environment.

Chapter 15
Digital Transformation of Springfield FC

15.1 The Fan Base

Our fans are at the heart of everything we do at Springfield Football Club (SFC). They have a deep permanent emotional attachment to the club, and we do all we can to offer them passionate sporting experiences with the team, with the players, and with fellow supporters.

SFC isn't just about the 90 min of the game, it's about being with our loyal fans all day, every day of the year. It's not just about the tens of thousands of fans who turn up on match day, it's also about the hundreds of thousands of fans who don't make it to the stadium, but follow the team from home, or wherever they are. And as we internationalise, it will be about our millions of fans worldwide.

Going beyond football to other aspects of our fans' daily lives, as a responsible member of the local community we're proud to have a registered charity that provides support in areas such as employability, education, health, inclusion and participation.

15.2 Omni-channel Multi-service

We've come a long way with digital. We've come a long way from the early days of having many separate channels to reach our customers. Nowadays, fans want to interact with us in the same way at many different times, through different devices, applications and channels. They follow us on Facebook, Twitter, YouTube, Instagram, SnapChat and Spotify.

We now have an omni-channel strategy that integrates the different choices available to consumers. So fans don't see a difference between our stores, our website, mobile, social and other channels. Fans can buy and consume experiences during matches in the stadium, in our Club Shop at the stadium, in the Club's outlet in town, on our Club website, on our news channel, online from their home computer, desktop

J. Stark, *Digital Transformation of Industry*, Decision Engineering, https://doi.org/10.1007/978-3-030-41001-8_15

or tablet, and from their smart phone when mobile. An important detail, whichever way fans choose to access, they use the same user name and password.

Fans can select and order merchandise whenever they want, from wherever they are. We're there for them on a 24/365 basis, with custom content and offers. They can select their favourite place in the stadium and order tickets whenever they want, from wherever they are. They can consume content, be it data, text, photo or video, wherever they are. It's all shared across social media platforms to maximise reach and engagement. Fans can interact with us and with each other through social media whenever they want.

Fans can follow all our games throughout the season in the stadium or online. They can watch our pre-season friendlies and see video clips of our training sessions. They can see video highlights and full match replays of our games. They can listen to interviews of our players and staff. They can read articles about the team and their favourite players in our Game Magazine and on our website. Fans can become Club Loyalty Members and get additional benefits and premium services. They can join through the membership scheme that serves them best. For example, the Adult scheme gives early access to tickets, hospitality facilities and discount vouchers. The Youth scheme includes a free team shirt, and entry vouchers to local family attractions.

Fans can look for, find and buy tickets online. They can browse and buy merchandise in store and online. This includes jerseys, shorts and socks for our Home Kit, Away Kit, Third Kit and Replica Kit. Everything's available in twenty sizes so they get a perfect fit. The Training Kit and the Travel Kit are also available. Not just for the first team, but also for the women's team, the Under-23s and the Under-18s. Our stores offer everything for the passionate fan. There are emblazoned balls, gloves, shin pads, towels, caps, scarves, car mats, window stickers, car toys, air fresheners, wallets, key rings, pens, pencils, notebooks, water bottles, dog coats, mugs, cards, phone cases, T-shirts, mouse mats, ties, cuff links, hot water bottles, sheets, pillow cases, bath robes, duvets, blankets, cushions, pyjamas, place mats, bottle openers, fridge magnets, etc. Everything a fan needs.

Supporting our omni-channel strategy is our in-house media team. Nowadays, our media operations are seen positively, not as an unavoidable cost centre. Our media team has a wide range of activities, including generating great content for our fans such as online catalogues for the store, news reports, studio discussions, interviews, statistics and heat-maps of players. As well as providing content for our website and apps, and the traditional newspapers, radio and television, our media team is working with an increasing number of social platforms. To support our growing global community of supporters we need to be on platforms not just in Western Europe and the US, but also in China, India, Japan, Latin America and Eastern Europe. Supporters in different geographies prefer different experiences, so our team customises the content to each geography and culture.

15.3 Multiple Revenue Streams

Just as there have been changes to our channels, the source of our revenues has also changed and diversified. Revenues no longer come just from ticketing for the 90 min of the game. Presently, they're from many sources such as media rights, transfer fees, merchandise, catering, hospitality packages, corporate partnerships, sponsorships, and advertising.

15.4 Fans as Profiled Consumers

Potentially our biggest source of revenues is our fan base and the associated huge volume of fan data. We used to have a patchwork of disparate platforms for ticketing, hospitality, shopping, and stadium entry that were unconnected with our main website. They had grown up separately over time, which made it difficult to track and understand supporter journeys and propose relevant products and services. Viewed individually, those legacy systems did a great job. Viewed together, they were holding us back. We realised we needed to change our approach, and put in place a solution that supported our business processes and maximised the value of our fan data.

Online, to serve our fans best, we're watching them on a 24/365 basis. Our analytic solutions consolidate fan data from multiple sources, then slice and dice it to uncover key characteristics of each fan. For example, we know what they like to buy, which players they like, what they like to wear and eat, their size, their age and sex, where they sit in the stadium, which video clips they watch, where they live, the members of their household, their interests outside football, and so on.

From the vast amounts of data, we build a personalised profile of each fan, a Digital Twin. With that knowledge, we aim to provide increasingly rich and relevant content and offers, seamlessly tailored to each individual fan, focused on their every need and desire. We're aiming to personalise products and services, for examples clothes with their name and that of their favourite player, and invitations to meet their heroes.

We see social media as helping us build the brand and build our audience. As we build the brand and the audience we attract more corporate partners, sponsors and advertisers. In parallel, we're looking to drive purchases to our sites and apps, maximising revenues from all that valuable data. In addition, we will also be able to directly monetise our prized huge volume of anonymised data to create a new revenue stream.

15.5   Connected Players

Many types of technology are coming into the game. For example, wearable technologies are giving us great data about player performance on the pitch. They provide huge quantities of data. We analyse this in different ways.

Player performance measurement gives us data about player positions on the pitch, their speed, their direction, when they have the ball, when they don't, the force with which they kick the ball, who they pass to, and so on. This data can be analysed in multiple ways, telling us about individual performance, collaborative performance and competitive performance. We can see where and how players perform best, as well as identify weaknesses to correct. We also track players' daily activities off the field, what they're eating, how they're sleeping and exercising. From data on opposing teams, we can work out and apply strategies where we are strong and they are weak. For example we may see that one player on an opposing team has a tendency to aim for a particular spot when taking a penalty. Another player may preferentially pass to one particular teammate or in one particular direction in certain situations. All this data and analysis helps us prepare for upcoming games and victories. In the future, we're looking to go further and thoroughly simulate games in depth digitally to test out and confirm our strategies.

15.6   Transfer Analytics

Just as we use analytics to scrunch through data on our players and our team with the objective of improving player and team performance, we also use it to identify any weaknesses in our team with a view to eliminating them.

Once we've identified a weakness in our team, we sift through the data on players of other teams, looking for players who would help us overcome our weak points. Once we've found a possible match, we dive deeper to see how the new player would fit with our players and team. The insights this generates are fed into our recruitment and transfer policies. They can help us find great new players. Equally, they can help us avoid recruiting players who may seem suitable at first glance, but wouldn't fit, for one reason or another, into our club environment.

15.7   In-Stadium Experience

Matches in our stadium are an important part of the fan's experience. This is another area where we're looking to improve. Currently this asset, with its great atmosphere, is digitally under-exploited. It's an area where there's little apart from replays of match incidents on large screens.

We're aiming to take a step forward, leveraging content related to the match, getting more information to fans, providing more close-ups and replays, increasing social-media activity, making insightful professional commentary available, introducing special "at the game" apps, bringing more attention and value to dynamic advertising boards, introducing more activities that involve the crowd—voting for man (or woman) of the match, best goal, best game photo, with rewards of seat upgrades, invitations to the Chairman's Lounge and the player's changing rooms.

We'd also like to share player performance data and analysis with our fans during the game, enriching their experience and understanding. We're trialling Point of View technology. This will allow fans to follow the game from the viewpoint of their favourite player on the pitch.

15.8 Connected Fans

We're expecting all the sensor and analytic technology currently being used for connected players to become available for fans.

This will open up new possibilities for fans. One possibility will be for them to compare their performance with that of their favourite players. Another possibility will be for fans to learn from their favourite players and improve their own performance. Another possibility, using Point of View technology and Augmented Reality technology, is to let them experience the game as if they were the player, really becoming part of the game. They'll be able to see the game through the eyes of the player and share their emotions.

15.9 Fantasy Football

Many fans love fantasy football games and enjoy displaying the skills they've built up over the years of supporting their team and studying the game. They know a lot about players and teams, and how they perform. We're looking for ways to benefits from this knowledge, such as crowd-sourcing for team selection.

We're expecting fantasy football to continue to grow with more and more player data being available for analysis. We're looking at ways to digitally play our team against opponents in fantasy virtual games, based on their real-world performance data. Fans will get even more involved, influencing player decisions. With digital technology, the sky's the limit for SFC. Players and fans will all be over the moon.

Chapter 16
Digital Transformation of TCBMC

16.1 Introduction

The Coach and Bus Manufacturing Company (TCBMC) manufactures coaches and buses. For several years, it's had the usual four in-house IT enterprise applications: Customer Relationship Management (CRM); Enterprise Resource Planning (ERP); Product Lifecycle Management (PLM); and Supply Chain Management (SCM).

TCBMC has a webmaster, and works with a Digital Agency. It has Digital Marketing and e-commerce specialists, an online shop and an online sales configurator.

Now the CIO is taking advantage of Cloud technology, using IT services offered by a Cloud provider over the Internet. That reduces the workload compared to installing and managing TCBMC's applications, servers, storage and databases. TCBMC is also taking advantage of mobile technologies that allow users to connect a device to the Internet and communicate from wherever they are. With mobile technologies, employees and customers don't have to be in a particular place when they communicate. So, for example, customers can access information from anywhere at any time. They can place orders online from anywhere at any time. People working in the company can access customer records, best practices, and eLearning courses online from anywhere at any time.

Looking to the future, the CIO created a Digital Transformation Team, and launched a benchmarking activity to better understand Digital Transformation. The other companies participating in the benchmark are seen as leaders in some aspect of Digital Transformation. They include companies from the fashion, consumer electronics, medical device, investment banking, and Fast Moving Consumer Goods (FMCG) sectors.

16.2 Smart Vehicles

The Digital Transformation Team has suggested that TCBMC should make its coaches and buses "Smart vehicles". This is seen as a first step towards making autonomous vehicles.

In this first step, TCBMC would equip its vehicles with electronic devices that make them "smart". Devices such as activators; actuators; displays; GPS receivers; memory; meters; microphones; motors; power modules; processors; receivers; Radio-frequency identification (RFID) chips; sensors; storage; switches; thermostats; transceivers, transmitters, vision systems and voice synthesisers.

Each type of device has its role in making a vehicle smart. Voice synthesisers "speak". A GPS device locates the vehicle. Displays show information. Microprocessors "think" and "calculate". Memories "remember" information. Receivers can receive information over a network. Transceivers and transmitters can communicate information over a network. Different sensors sense and measure different things, such as: acceleration; humidity; light; location; movement; pressure; proximity; sound; speed; stress; temperature; touch and vibration.

The Digital Transformation Team says that with smart vehicles, TCBMC can offer customers new products. And it will be possible to enhance existing vehicles with new functions and features. For example, a smart bus has sensors that can see and avoid any obstacles in its way, measure the road surface conditions, count the number of passengers, know its location, measure and record performance parameters, and remember its maintenance history.

The vehicles will be able to read their own sensor data, and check that they're in good shape. If they're not, they'll be able to make real-time adjustments to their behaviour based on their sensor data. If they see that they're not being used in the best way, they'll be able to inform drivers and maintenance staff. And when they get to the end of their lives, full information on their history will be available, enabling selective component reuse and recycling. This information will be used: to predict remaining component lifetimes; to optimise recovery; to process components and assemblies effectively; and to reduce environmental impact.

Connecting a smart vehicle to a communications network makes it a Connected Product. When a vehicle is connected to the Internet, it becomes part of the Internet of Things. Once vehicles are connected to a communications network, control signals can be sent to control them. And the data that the vehicles generate can be communicated and captured. The vehicle can send information (the Voice of the Product) about its performance back to the user and/or to TCBMC. This information is product data. It's Big Data from the use/support phase of the product lifecycle. For example, a smart coach can send a message to the coach operator when it needs maintenance, and it can send a message to TCBMC saying which parts were over-engineered.

The Digital Transformation Team claims that the sensors on a coach could generate a terabyte of sensor data on a long distance trip. This data is part of Big Data. TCBMC also gets social Big Data. For example, from the trillion online searches made each year. It wants to know who's searching, where it's being "liked", and if it's being mentioned by top social influencers. Within all the Big Data generated each day, there's valuable information about TCBMC's vehicles and activities. TCBMC would like to know that information. However, so much Big Data is produced, that it would take a human being many years to read and make sense of it. Instead, that role is taken by Analytics.

Analytics includes the computerised analysis, or "mining", of huge volumes of data; predictions made based on the analysis; decisions taken as a result of the analysis; and value added as a result of the predictions, decisions and/or analysis. The algorithms analysing the data search for correlations, patterns and meaning. Their findings can be transmitted back to the vehicle operator or to TCBMC, or other participants in the product lifecycle. The information could be used in many ways. For example, it could be used to: better understand customer behaviour; identify and define new high-value services; tailor vehicles to meet the desires of individual customers; and target existing customers with add-on services. Analytics could also help TCBMC to: understand how vehicles are being used and behave; monitor vehicle degradation; replace components before failure, rather than after failure; monitor vehicle quality; customise support; and provide customers with intelligent recommendations about vehicle use. The information will also be used to help develop new ideas for vehicles, and to develop future vehicles.

16.3 Industrial Internet of Things

The Digital Transformation Team investigated the Industrial Internet of Things (IIoT). This is similar to the Internet of Things, but is limited in scope to the company's manufacturing environment. As a result, the "things" of the Industrial Internet of Things are equipment in the factory such as a scanner of incoming goods, a machine tool, a robot, and a test rig. In the factory, the IIoT enables real-time data collection, monitoring, display and controlling of production resources. All the equipment, new and existing, will be equipped with electronic devices and connected to a communications network. Electronic devices will sense factory equipment at a very high rate. The sensors on the machine could be read many times a second. With many smart machines in a factory, the factory would then generate many megabytes of data per minute. Data will be collected to monitor real-time performance. Analysis of the data will improve activities and decision making. For example, the risk of machine breakdown will be reduced by scheduling preventive maintenance based on real-time data. Robots will learn new tasks while working on the job. Machine settings for the next product will be simulated before the physical changeover, reducing setup time.

16.4 Intelligent Automated Factory

The Digital Transformation Team sees possibilities for TCBMC to aim for an intelligent automated factory. The machines will be smart. They'll be fitted with electronic devices such as sensors, processors and transmitters. With these devices, the machines will do more than basic manufacturing functions. The devices on the machines will measure characteristics such as movement, strain, temperature and vibration. They'll see what's been produced, and control the quality. Then they'll report performance. Data reported by the machines will be aggregated to create high value information. And used to take decisions about what to do next. As the factory will be intelligent, it will be possible to reconfigure and re-purpose intelligent equipment to respond quickly to changes in customer demand. Robots will become autonomous. They'll work collaboratively with people. They'll help people by doing the heavy lifting and repetitive work. They'll be able to self-optimise and take their own decisions.

16.5 Connected, Integrated Factory

The Digital Transformation Team proposes a connected, integrated factory environment for TCBMC. Shop floor data will flow freely through the communications network. There'll be vertical integration from shop floor sensors up through Manufacturing Execution Systems (MES) to the corporate Manufacturing Resource Planning system. Horizontally, the factory systems will be connected and integrated to other systems in the company such as Customer Relationship Management (CRM). As appropriate, the factory systems will connect through to customer and supplier systems.

16.6 Digital Factory

The Digital Transformation Team recommends that TCBMC aims for a digital factory. Digital technology will be used to run the manufacturing process. The company will have a clear data picture of the factory. This will show what's really going on. There'll be real-time visibility into the key parameters of process and product variance.

16.7 Augmented Reality Factory

The Digital Transformation Team contemplates the possibility of a TCBMC Augmented Reality (AR) factory. Shop floor workers will have an augmented view of the factory with computer-generated images overlaid on their work environment. The augmented view will give them a better understanding of the situation. And allow them to take better decisions. Assembly workers will be assisted by work instructions projected on the parts they're assembling. Wiring installers will be able to fly through 3-dimensional virtual models to help install electrical wiring.

16.8 Analytic Factory

TCBMC's Digital Transformation Team wants an analytic factory. Big Data from the smart, connected equipment and from other sources will be analysed. Analytics will support many activities, for example to: monitor product quality; identify process bottle-necks; detect defects; identify likely production problems and prevent them happening; foresee machine and tool wear and tear; detect potential failure points on the shop floor; launch preventive equipment maintenance; improve energy efficiency; minimise downtime; and reduce operating costs.

16.9 Take-Charge Factory

The Digital Transformation Team counsels a take-charge factory. The aim is to predict what's going to happen before it happens. As a result, for example, it will be possible to carry out preventive maintenance rather than wait for reactive maintenance after a problem has occurred. Equipment will be simulated and optimised before use.

Simulation is low-cost, fast and effective. Mathematical modelling enables simulation of performance before physically installing equipment. The performance of equipment will be studied before it's been physically built or implemented. It will be possible to see how everything works before installing any equipment. Simulation will use the 3D as-designed model of the factory environment. Factory digitalisation will generate a three-dimensional image of the as-installed factory environment. Later, these digital factory twins will be used when making changes and calculations at the factory.

16.10   Adaptable, Flexible Factory

The Digital Transformation Team advises aiming for an adaptable, flexible factory. As customer demand changes, it will be possible to make changes nimbly and scalably in production.

16.11   Secure Protected Factory

The Digital Transformation Team call for a secure protected factory. Security policies will be defined, documented and communicated so that everyone is aware of them. Access to the factory will be controlled with key cards, biometric techniques and video surveillance. Production will be kept running at all times with an uninterruptible power supply. Cyber-security will be implemented to protect against hackers, thieves, industrial spies, disgruntled customers and employees, and other cyber attackers.

16.12   AI-Augmented Factory

The Digital Transformation Team asserts the need for TCBMC to aim for an Artificial Intelligence (AI)-augmented factory. AI will be used to help respond to changing situations with the correct decisions. Cognitive applications will enhance people's natural abilities, skills and knowledge. AI will support shop-floor workers throughout the factory with trustworthy recommendations.

16.13   Advanced Manufacturing Factory

The Digital Transformation Team advises use by TCBMC of new manufacturing technologies, such as additive manufacturing, on the shop floor. 3D printers will be used to produce parts directly from CAD models. With 3D printers, it will be possible to make parts that are difficult to make with traditional manufacturing techniques, and to repair these, and other, parts.

16.14   New Business Processes

The Digital Transformation Team expects TCBMC will need to define or redefine many business processes to benefit from new technologies and opportunities. Some processes were defined years ago, but need to be looked at again. And then there are other processes the company didn't have before. The Digital Transformation Team says the company should look at its processes for managing the product portfolio, for visioning, for road-mapping, and for product planning. A process will be needed to measure the value of the product portfolio. The company should look at the ideation process. New processes are needed to manage the increasing amount of electronics and software in products. TCBMC will also need a process to handle Big Data, and new processes for managing customer requirements. It also needs updated processes in the areas of security, intelligence, collaboration and intellectual property.

16.15 Skill Gap and Training

The Digital Transformation Team recognise that, due to the emergence of new technologies and changes in the product environment, there are likely to be areas where TCBMC has significant skill gaps. The required new roles will need to be identified. People with appropriate skills, knowledge and experience will be hired. For example, due to new technologies, TCBMC will need to hire specialists in areas such as Electronic CAD (ECAD), Application Lifecycle Management (ALM), blockchain, AI and IoT Platforms. As well as people who can use new technologies, it will also need people who can understand how to take advantage of new technologies, and people who can implement and integrate them in the company.

The Digital Transformation Team expects a significant investment in training. Training will convey basic topics, lessons learned, individual procedures and methods, as well as address technology and systems. Employees can expect a lifetime of learning. They'll need training in the skills they're lacking, new technologies, advances in existing technologies, new versions of the software the company uses, and in the company values such as environmental respect, ethics, integrity, sustainability and human rights. Everyone will learn about the business processes. It's important they know who does what, what has to be done, where everything is, and how things are organised. With mobile technologies available, people can learn anywhere, at any time. Training will be enhanced by use of Virtual Reality and Augmented Reality technologies.

16.16 Digital Transformation Team Report

The Digital Transformation Team summarised its finding in a report and presentation. The Team leader presented these to the CIO, who was excited by the many possibilities, and in turn, presented them to TCBMC's Chief Financial Officer (CFO).

16.17   CFO Reaction

After the presentation, the CFO congratulated the CIO on the report and the work done by the Digital Transformation Team. However, she said that it wouldn't be accepted by the CEO in its current state.

She reminded the CIO that the CEO was under great pressure from the President to improve performance. She showed the CIO some slides from the final presentation that had recently been made by world-class strategy consultants after reviewing TCBMC's performance. She pointed to one slide showing some findings about recalls of products, deaths resulting from product use, and penalty payments to customers. Another slide addressed a lack of innovation, poor product design, long product development times, unauthorised changes made to products, poor lifetime support of products, and no end-of-life strategy for products. Another slide compared the company to a government bureaucracy, everything being slow, no-one taking decisions, a multitude of committees and meetings, departments not working together.

The CFO told the CIO that the CEO was under daily pressure to solve these issues, and was looking for solutions. The CIO suggested another Cost Reduction program, but that was turned down by the CFO. She went on to say that as so many of the problems were related to products, the solution would need to have a focus in that direction.

She said that one solution being looked at by the strategy consultants was Product Lifecycle Management. She asked if the CIO thought that could link to Digital Transformation. The CIO said they had a Product Data Management solution and that could fit.

The CFO hesitated. She said she had a slide pack about Product Lifecycle Management from the strategy consultants. She said she'd show the CIO some of the slides. The first slide was titled Product Lifecycle Management (PLM).

16.18   The CFO Presents PLM

The CFO showed the next slide. It was titled the PLM Grid, and showed a 5 * 10 matrix. Apparently it showed the scope of PLM. On the horizontal axis were the five phases of the product lifecycle: ideation; definition; realisation; use and support; disposal and retirement [1].

On the vertical axis of the grid were the ten components that have to be addressed when managing a product across the lifecycle. The ten components were: business processes; product data; the Product Data Management (PDM) system; other PLM applications; people; facilities and equipment; product-related methods and techniques; products; management and organisation; and objectives and Key Performance Indicators (KPIs). The PLM Grid appeared to cover all the product-related areas the consultants talked about.

On the next slide, just to be clear about the scope, the strategy consultants had listed some of the product-related activities of a company that fit under the PLM umbrella: "managing a well-structured and valuable Product Portfolio; maximising the financial return from the Product Portfolio; managing products across the lifecycle; managing product innovation, development, support and disposal projects effectively; providing control and visibility over products throughout the lifecycle; managing feedback about products from customers, products, field engineers and the market; effectively managing product requirements; enabling collaborative work with design and supply chain partners, and with customers; managing product-related processes so that they are coherent, joined-up, effective and lean; capturing, securely managing, and maintaining the integrity of product definition information; making product definition information available where it's needed, when it's needed; and knowing the exact characteristics, both technical and financial, of a product throughout its lifecycle."

Then the CFO read out a definition from another slide, "*Product Lifecycle Management (PLM) is the business activity of managing, in the most effective way, a company's products all the way across their lifecycles; from the very first idea for a product all the way through until it is retired and disposed of. PLM is the management system for a company's products. It manages, in an integrated way, all of its parts and products, and the product portfolio.*"

"And", she said, "Here's the objective. *At the highest level, the objective of PLM is to increase product revenues, reduce product-related costs, maximise the value of the product portfolio, and maximise the value of current and future products for both customers and shareholders.*"

The CFO paused and said, "It looks to me like something we need. What do you think?"

"Looks great to me. It addresses all the problem areas the strategy consultants highlighted", replied the CIO, "I'll get the Digital Transformation Team to look in detail."

"Great," said the CFO. "Check it out. Things are moving fast. Let's meet again at the end of the week."

Reference

1. Stark J (2019) Product lifecycle management (volume 1): 21st century paradigm for product realisation. Springer. ISBN 978-3030288631

Chapter 17
Select Digital Technology

17.1 Recap

After reading about the Digital Transformation (DT) activities of organisations in a variety of industry sectors in previous chapters, it's useful to recap some of the key points from earlier chapters of the book.

17.1.1 Definition

In this book, which addresses the digital transformation of industry, the following definition of digital transformation is used.

Digital Transformation is the transformation of part or all of an industrial organisation, through the application of a particular digital technology, or technologies, to improve one or more of its activities.

17.1.2 Digital Technologies

There are many digital technologies that could be applied to achieve Digital Transformation. They include: Additive Manufacturing; analytics; apps; Artificial Intelligence; Augmented Reality; automation; autonomous vehicles; Big Data; blockchain; blogs; Cloud computing; database technology; e-commerce; the Internet; the Internet of Things; Knowledge Management; machine learning; mobile technology; robotics; smart connected products; smart phones; social technology; streaming; telecommunications; Virtual Reality; the World Wide Web; and websites.

17.1.3   Business Processes and Activities

In an industrial organisation, there are many business processes to which Digital Transformation can be applied. They include: Management processes (such as Governance, Quality, Financial Planning, and Merger and Acquisition); Support processes (such as Human Resources, Finance and Administration, and IT); and Operational processes (such as Supply Chain Management, Customer Relationship Management, and Product Lifecycle Management).

17.2   Difference Organisation, Different Case

Previous chapters have shown many differences between the Digital Transformation activities of organisations in different industry sectors. For example, there are differences in the digital technology they have already implemented and are using. And, as the following sections show, there are also many other differences.

17.2.1   Different Activities, Different Resources

There is a lot of commonality between the organisations described in the previous chapters, even though they are in different industry sectors.

However, there are also many differences between these organisations because they also carry out some very different activities and make use of different resources.

17.2.2   Different Current Situations

There are differences in the current situations of different organisations. In addition to their different activities and resources, they are also different in many other ways. For example, they have different objectives, different maturity, different knowledge levels, different viewpoints, different business strategies, different financial situations, and so on.

17.2.3   Different IS Maturity

There are differences in the IS maturities of the organisations. Some have little IS experience. Others are, from an IS view, very mature.

17.2.4   Different IS Strategies

There are differences in the existing IS strategies of the organisations. These are likely to lead to different future choices of digital technology.

17.2.5   Different Objectives

There are differences in the objectives of different organisations. Some have clearly defined objectives, others don't. Some have objectives related to costs, others have objectives related to revenues. These differences are likely to lead to different future choices of digital technology.

17.2.6   Different Viewpoints

There are differences in the viewpoints of different organisations. Some look at Digital Transformation from a technology viewpoint, others from a business viewpoint.

17.2.7   Different Scopes

There are differences in the scopes of the Digital Transformation activities of different organisations. Some only see new technologies in their scope. Some also see changes to the way that the organisation will operate. And some also see changes to the way that individuals will work.

17.2.8   Different Knowledge Levels

There are differences in the Digital Transformation knowledge levels of different organisations. Some are very knowledgeable, others know little about Digital Transformation.

17.2.9   Different Levels of Involvement

There are differences in the levels of involvement of different organisations. Some only have executive involvement in Digital Transformation, some have executive and technical involvement, and some only have technical involvement.

17.2.10   Different Levels of Detail

There are differences in the level of detail addressed in different organisations. Some organisations aim to understand as much as possible in great detail before moving ahead. Others prefer not to go into detail before selecting specific technologies.

17.2.11   Different Financial Situations

There are differences in the financial situations of different organisations. Some have strong financial constraints. Others are less constrained, but still need to justify any investments.

17.2.12   Different Business Strategies

There are differences in the business strategies of different organisations. Some are aiming to be leaders in their sector. Others are followers. Such differences are likely to lead to different future strategies for Digital Transformation.

17.2.13   Different IS Roles

There are differences in the relative importance of IS in different organisations. Some see IS as a key competence to be strengthened. Others see IS as secondary, and give priority to strengthening other competences.

17.2.14 · Different Benefits to Achieve

There are differences in the benefits that different organisations want to achieve with Digital Transformation. Some want to solve existing problems. Others want to seize new opportunities.

17.2.15 Different Positions on the Journey

There are differences in the positions in which different organisations find themselves on their Digital Transformation journeys. Some have not yet started their Digital Transformation. Some are just starting out on Digital Transformation. Others are already some way down their Digital Transformation path.

17.2.16 Different Approaches (1)

There are differences in the approaches of different organisations. Some start by reviewing their mission, or the way that their business has already changed. Some talk to customers to find out their views. Other start by assuming that they need to apply a particular digital technology.

17.2.17 Different Approaches (2)

Another difference in the approaches of different organisations is that some see Digital Transformation as a journey. Others see it simply as the implementation of a new technology.

17.2.18 Different Approaches (3)

Another difference in the approaches of different organisations is that some see Digital Transformation simply as the implementation of a new technology. Whereas others see a Digital Transformation Program as an activity that changes the organisation.

17.2.19   Different Approaches (4)

Another difference between organisations is their willingness to share information about Digital Transformation with other organisations in the same industry sector. In some sectors, sharing best practices is seen as a benefit for all organisations in that sector. In other sectors, best practices are seen as a source of competitive advantage that should not be shared.

17.2.20   Different Team Structures

There are differences in the types and organisations of people involved in Digital Transformation activities in different organisations. In some cases, people are loosely involved in Digital Transformation activities. In other cases, Digital Transformation teams have been formally set up and structured. Some teams only include IS people, others have a mixture of IS people and business people.

17.2.21   Different Affected Areas

There are differences in the areas of the organisation that are targeted for change. Some organisations target change across the entire organisation. Others are only looking to change in one very specific area.

17.2.22   Different Technologies Selected

There are differences in the technologies that organisations select.

There are many digital technologies that an organisation could select to achieve Digital Transformation. These technologies include: Additive Manufacturing; analytics; apps; Artificial Intelligence; Augmented Reality; automation; autonomous vehicles; Big Data; blockchain; blogs; Cloud computing; database technology; e-commerce; the Internet; the Internet of Things; Knowledge Management; machine learning; mobile technology; robotics; smart connected products; smartphones; social technology; streaming; telecommunications; Virtual Reality; the World Wide Web; and websites.

Most organisations choose several of these technologies. But they don't select exactly the same technologies. They select the technologies that correspond to their objectives, and fit to their investment possibilities.

17.3   Need for an Organisation-Specific DT Approach

There are so many differences between organisations that it wouldn't be appropriate to apply the same Digital Transformation approach in them all. Instead, each organisation has to find an approach that fits its needs and circumstances.

17.4   Need for a Digital Transformation Program

The Digital Transformation of an organisation is likely to affect many of its functions. It's also likely to be expected to lead to benefits in many functions.

Digital Transformation is likely to involve many activities such as defining objectives, understanding digital technologies, carrying out a Feasibility Study, understanding the "as-is" situation, defining the "to-be" situation, identifying costs, justifying an investment, planning, training and implementing.

Digital Transformation is likely to impact many of the organisation's functions such as Marketing, Sales, Research, Engineering, Logistics, Service, Finance, Quality, Human Resources and Information Systems (IS).

With such a wide-ranging scope, Digital Transformation should be addressed with a cross-functional organisation-wide program. This program will be specific to the particular organisation.

17.5   Digital Technology Selection

Selection of Digital Technology is unlikely to be the first activity in a Digital Transformation Program. Most likely it will be preceded by many other activities such as defining the Program's objectives, setting up a Digital Transformation Team, identifying and documenting the expected benefits of the Program, identifying the costs of candidate digital technologies, comparing the costs and benefits of different approaches, justifying the investment for the Program, etc. These activities are addressed in the following chapters.

Chapter 18
Run a Program

18.1 Introduction

The Digital Transformation (DT) of an organisation may affect many of its functions such as Portfolio Management, Product Management, Marketing, Sales, Research, Engineering, Regulatory, Logistics, Production, Service, Recycling, Finance, Quality, Business Process Management, Human Resources and Information Systems (IS). As a result, Digital Transformation should be addressed with a cross-functional organisation-wide Digital Transformation Program, run with a Project Management approach.

Digital Transformation shouldn't be run as a purely IS undertaking just involving people from the IS or CIO function.

18.2 Importance of a Project Management Approach

A project is a temporary activity carried out by an organisation to achieve a specific goal. It has an intended start date and an intended end date. A project differs from most activities in an organisation. Most activities aren't temporary. They're permanent, everyday, routine, and organised to achieve the same goal many times [1].

For example, one of my customers was a company that had a manufacturing plant which produced motors for the transportation industry. Every day, about a dozen motors came off the end of the line. The plant carried out the same tasks every day for several years. The workers in the plant didn't do project work, they did everyday routine work. In a company's everyday routine operations, such as those in that manufacturing plant, roles and responsibilities are clear. They're described in departmental guidelines. People work in a particular department such as Manufacturing. They work the way their departmental boss, such as the Manufacturing Plant Manager, or the Manufacturing VP, wants them to work [2].

J. Stark, *Digital Transformation of Industry*, Decision Engineering,
https://doi.org/10.1007/978-3-030-41001-8_18

However, in most projects, the situation is different. The project is unique and cross-functional. It's run once, it's not repeated time and time again. Because it includes people from several functions, it's not clear who should be the boss. To address these issues, Project Management, an approach to management that is specific to projects, is needed.

Project Management is the business process for managing projects. The objective of Project Management is to achieve the project's goal on-time and within the project budget. Project Management is made up of activities such as planning, scheduling, organising, allocating, leading and controlling of company resources.

A Project Manager is responsible for on-time, on-budget achievement of the project's goals.

A Project Management Office (PMO) may provide administrative support to the Project Manager and to other project participants (Fig. 18.1). The role of a PMO varies widely from one project to another. In some projects, the PMO is a key hands-on participant in managing the project. In other projects, the PMO may just have an advisory role. And sometimes a project doesn't have a PMO.

18.3   Program

Just like a project, a program is a temporary activity carried out by a company to achieve a specific goal [3].

However, a program is made up of many related projects. Some of these may run in parallel, some may run in series. All of the projects contribute to achieving the goals of the program. Each project within the program may start and end at a different date (Fig. 18.2).

advising on project management best practices	organising training
preparing and maintaining the project schedule	organising meetings
giving training on special techniques	tracking costs
tracking project progress	preparing risk analysis
preparing for issue analysis	preparing project plans

Fig. 18.1   Examples of support provided by a Project Management Office

Timescale	Project A	Project B	Project C	Project D
Y1 Q1	Phase 1			
Y1 Q2	Phase 2	Phase 1		
Y1 Q3	Phase 3	Phase 2	Phase 1, Phase 2	
Y1 Q4	Phase 3	Phase 3	Phase 3, Phase 4	
Y2 Q1	Phase 4	Phase 3		
Y2 Q2		Phase 4		
Y2 Q3				Phase 1, Phase 2
Y2 Q4				Phase 3, Phase 4

Fig. 18.2   Projects in a program

In most Digital Transformation Programs, there will be many projects and sub-projects. There are many advantages of grouping related projects in a program (Fig. 18.3).

A Program Management Office (PMO) is similar to a Project Management Office except that, instead of addressing a single project, it addresses the many projects of a program.

18.4 Project Schedule, Program Schedule

A project schedule is a list of a project's tasks that also shows other information, including intended start and finish dates for each task. Examples of the other information it may show include the name of the person responsible for the task, the resources assigned to the task, and the tasks' deliverables. There are many reasons for making a project schedule (Fig. 18.4).

18.5 Risk of Program Failure

Digital Transformation Programs have many of the characteristics that typify, in other domains, projects that have a high risk of failure. For example, they have cross-functional aspects, lead to organisational change, and involve changing the way that people work. Most projects in a Digital Transformation Program will be cross-functional and include people from different backgrounds who may have very different understandings of Digital Transformation. Some of the people in the Program may know little about Project Management. Some members of the Program Team may not have participated before in cross-functional activities. In such circumstances, it can be difficult for them to know how best to go about their work.

more effective use of resources	easier to report all the projects in the program
easier to plan all the projects in the program	decision-making is clearer and more effective
easier to manage all the projects in the program	likelihood of success is increased
easier to control progress of the projects	less waste, shorter time frames and lower costs

Fig. 18.3 Advantages of a program

identify interactions between tasks	link deliverables to tasks in the project
understand interactions between tasks	communicate the use of resources
show interactions between tasks	reduce uncertainty
better understand the objectives	clarify timelines of tasks in the project
create a basis for monitoring progress	understand workloads of participants in the project

Fig. 18.4 Reasons for making a project schedule

Although the project management of all cross-functional projects is challenging, Digital Transformation Programs have some special characteristics (Fig. 18.5) that can exacerbate the situation.

18.6   Types of Program Failure

There are many ways Digital Transformation Programs fail. Examples include failure to meet objectives, failure to deliver on promises, failure to keep within budget, failure to report meaningfully, failure to develop accurate specifications, and failure to get people to change their behaviour. Many of the reasons for failure are linked to poor project management.

Failure is rarely due to the quality of a particular new process or the functionality of a new digital solution. Usually it's due to the way that these are implemented, which in turn is a consequence of the way the project is managed. The likely consequences of poor Project Management in the Digital Transformation Program are serious (Fig. 18.6).

18.7   Causes of Failure

In a typical company, there may be many issues with projects in the DT Program (Fig. 18.7).

there can be many projects and sub-projects in a DT Program	DT Programs run over a long period
successful DT involves significant organisational change	DT Programs have a wide scope
generic Project Managers may be out of their depth in DT details	DT Programs address many issues

Fig. 18.5 Special characteristics of project management in DT

projects come in late	executives are unhappy
projects lead to incorrect results	Digital Transformation is seen as a failure
projects give partial results	Team Members resign or are fired
the Program Manager is fired	Team Members are reassigned

Fig. 18.6 Consequences of poor project management in the DT program

Fig. 18.7 Issues in the digital transformation program

Project Management guidelines ignored	scope creep
unclear scope and business need	poor communication
inappropriate support technology	unclear objectives
uncommitted sponsor and stakeholders	poor project start-up
expected benefits not clearly defined	overrun
insufficient team resources	lack of governance
benefits not achievable	team members run wild
weak business case	no business case

For example, project objectives may not be clearly defined. The result will be confusion. There is often duplication or overlap of activities between different projects. Boundaries between projects are rarely defined. Project Governance may not be clearly defined. Roles and responsibilities may be vague. As a result, nobody feels really responsible for the project. If any problems arise, no-one's there to put things right. There may be no Key Performance Indicators (KPIs) in some projects. In other cases, the wrong performance measures are used to measure progress.

Sometimes, the Project Management processes in a company haven't been defined. If they have, there may be no management commitment to ensure that they're followed. The processes may be poorly defined, and poorly documented with a resulting lack of clarity about what should be done. There may be no training about Project Management.

18.8 Overcoming Issues

Project Management offers features to help overcome any issues arising in the DT Program and achieve a successful Digital Transformation (Fig. 18.8).

18.9 Success Factors

For successful Digital Transformation, run a Digital Transformation Program. Take a project management approach.

Many activities in the Digital Transformation Program are likely to be challenging. Some success factors for these activities, drawn from experience in many Programs, are shown in Fig. 18.9. Understanding these can put, and keep, the Program on a good track.

leads to faster delivery	supports effective teamwork
enables use of common templates	reduces cost overrun
more predictable results	manages risks better

Fig. 18.8 Features of project management

clear Business Case and Objectives	motivated Program Team
good Program Sponsor	good Risk Management
good planning, good project schedule	early success
well defined roles and responsibilities	avoiding project creep
experienced Project Managers	appropriate KPIs
including 15% contingency in plans	common vocabulary
agreed acceptance and success criteria	learning from experience
good communication between participants	same Program Sponsor throughout
closing the Program with a clear cut-off	good grip of roles and responsibilities

Fig. 18.9 Success factors for project management in the DT environment

helps understand future activities	stops things slipping off the radar
shows milestones and deliverables	leads to more realistic time-scales
leads to more accurate costing	details resource requirements
gives early warning of slippage	keeps everyone aware of progress

Fig. 18.10 Benefits of a good project schedule

The benefits of a good project schedule can also be identified from experience in many DT Programs. Some examples are shown in Fig. 18.10. A good schedule is one of the most important components of a Digital Transformation Program.

References

1. Project Management Institute (2017) A guide to the project management body of knowledge. Project Management Institute. ISBN 978-1628251845
2. Stark J (2019) Product lifecycle management (volume 1): 21st century paradigm for product realisation. Springer. ISBN 978-3030288631
3. Subramanian S (2015) Transforming business with program management: integrating strategy, people, process, technology, structure, and measurement. Auerbach Publications. ISBN 978-1466590991

Chapter 19
Involve Your Executives

19.1 Reasons for Involving Executives

The Digital Transformation (DT) of an organisation is likely to involve many of its functions such as Portfolio Management, Product Management, Marketing, Sales, Research, Engineering, Regulatory, Logistics, Production, Service, Recycling, Finance, Quality, Business Process Management, Human Resources and Information Systems (IS). It may also involve customers and suppliers.

The program will address the long term. It will be related to future organisational activities and performance. It may have high costs. There'll be a need to prioritise activities. The only people who can start a cross-functional organisation-wide program of such a magnitude are executives. Only they have the strategic business focus, knowledge, experience, authority and responsibility to launch such a Program and ensure a fit with the organisation's requirements [1].

19.2 Early Involvement and a Long Haul

It's important to get executives involved early in the activities leading up to the Digital Transformation Program. The earlier they're involved and participate, the more they'll understand, and the more they'll be supportive later.

A Digital Transformation Program is a major undertaking. It's organisation-wide. It's going to change the way the organisation works. It's likely to meet resistance to change. It's a challenging program that will run for quite a few years. Full achievement is likely to take a lot of effort and a lot of time. It isn't going to happen overnight. It isn't realistic to expect everything to be done instantaneously. Executives should be prepared for the long haul.

J. Stark, *Digital Transformation of Industry*, Decision Engineering, https://doi.org/10.1007/978-3-030-41001-8_19

19.3 Executives Roles

What role should executives play in Digital Transformation? From my experience in many programs, executive roles in Digital Transformation Programs are similar to executive roles in any other activity [2]. In general, ten roles are expected of an organisation's executives (Fig. 19.1).

19.3.1 Maintain Awareness, Provide Vision

The first role of the executive is to maintain awareness and provide vision. Executives need to know where the organisation has been, where it is, and where it's going.

Executives are expected to know why the organisation exists, where it's been, and where it is now. This requires them to maintain constant awareness of the surrounding landscape, customers, technology developments, improvement opportunities, and so on.

Executives are also expected to know where the organisation is going, and why. They need to create and communicate the organisation's vision and overall direction.

19.3.2 Set Objectives and Values

Knowing where the organisation is, and where it needs to go, executives can set objectives and define core values. Objectives have to be defined in such a way that the corresponding outcomes can be measured. Otherwise, nobody would know if progress is being made, or if the objectives have been achieved.

Fig. 19.1 Ten executive roles

1	Maintain awareness, provide vision
2	Set objectives and values
3	Oversee governance
4	Lead
5	Represent and communicate
6	Ask questions, give answers
7	Identify and develop leaders
8	Monitor progress, measure outcomes
9	Take decisions and corresponding action
10	Hold accountable and provide recognition

19.3.3 Oversee Governance

The next role of the executive is to oversee organisational governance. The governance of an organisation includes all of its governing structures and processes. This includes designing the organisation in such a way that it meets its objectives and achieves its mission. It includes developing strategies and launching corresponding activities. It includes the way that actions, policies and roles are structured and applied. It includes the definition of roles and the relationships between them. It includes the supervisory and overseeing activities of making sure that the organisation operates as intended.

19.3.4 Lead

The next role of the executive is to lead. There are several facets to the executive role of leadership. One is to motivate and inspire. Another is to be a role model, to show the way, to lead by example. Another facet is to guide and direct the work of others. Other facets of the leadership role are to have a positive, enthusiastic, confident attitude and to launch key activities for the organisation's future.

19.3.5 Represent and Communicate

Another executive role is to be a figurehead, a spokesperson, a representative. An executive represents the organisation in many situations, both internally and externally.

Internally, executives may represent the organisation in situations with other executives, managers and employees.

Externally, executives may represent the organisation in situations with stakeholders, suppliers, industry organisations, local and national government, and so on.

The executive has a role to communicate with a wide range of people and entities, keeping them informed as appropriate.

19.3.6 Ask Questions, Give Answers

Executives have roles of asking questions and of providing answers.

Executives need to ask questions to find out what's happening. Indirectly, asking questions encourages others to keep aware. Good questions will help them see new ways of solving problems and making progress.

Executives need to give answers. Again, this can help people to move forward in specific circumstances. Giving answers can also reinforce understanding of vision, goals, and values.

19.3.7 Identify and Develop Leaders

Identifying and developing leaders is another key role of executives. Executives need to look for, and to find, more leaders to help the organisation achieve its current goals and to continuously succeed. Most of those found will need to be further developed. Executives need to put in place mechanisms to ensure continuous development of leadership skills.

19.3.8 Monitor Progress, Measure Outcomes

Executives have roles to set objectives and develop strategies, and to launch important activities for the organisation. They also have a role to monitor the progress and measure the outcomes of these activities. After activities have been carried out, the outcome needs to be measured to evaluate the success of the organisation in meeting objectives.

19.3.9 Take Decisions and Corresponding Action

Executives have roles to set objectives, launch activities and measure activity outcomes. Once the outcome is clear, executives need to decide what to do next. They need to take informed decisions quickly. They need to take appropriate action when the actual outcome differs from the requirement. And make sure it happens.

19.3.10 Hold Accountable and Provide Recognition

Executives have a role to hold the employees of an organisation accountable for their actions. This may be through reward and recognition, or through something less pleasant.

19.4 Key Executive Activities

The above roles apply in most types of organisation. They also apply in a Digital Transformation Program. For example, executives should take the lead on Digital Transformation and show the way. Nobody else in the organisation can launch, lead and maintain a major organisation-wide, cross-functional program such as Digital Transformation.

Executives should set the objectives of the program, and define the strategy. Nobody else can. They should track progress, and make sure the program stays on time, on budget and in scope. And, if things aren't working out, they should take action and put the program back on track.

Executives should target the Digital Organisation. Executives are responsible for the entire organisation, whereas the other people in an organisation are focused on a particular area. People focused on a particular area of an organisation will want to improve that area, what they understand, and what they'll be judged on. That may be their department, or their group, or section. But Digital Transformation is for the whole organisation, not just for one department. Executives should target a Digital Transformation Program that addresses the whole organisation.

Executives will be involved in a wide range of activities in a Digital Transformation Program. These can include program financing, setting goals, serving on steering committees that focus on specific problems or activities, reviewing progress, approving program deliverables, and being leaders.

Executives must make sure that the upfront planning for the Digital Transformation Program is done, documented and communicated. A lot of detailed planning is needed to define how the organisation will work in the future digitally transformed environment. This is important work. It's better to work out how the environment will look in the future, rather than just let things happen.

Executives should make sure that important Digital Transformation Program information is documented. Examples of such information include the Program Charter, the Vision, the performance indicators, and the plans. If this information is not documented and communicated, it could soon be forgotten and misrepresented. And then people won't be sure about what they're doing and what they should do next.

Executives need to set up the Digital Transformation Program governance so that the program's structure, roles and responsibilities are clear. That includes making sure there's a sponsor for the Digital Transformation Program, a Steering Committee, a Program Leader, a core Team, and perhaps other teams. Executives should make sure that the program is correctly organised and staffed. They should select a good, committed Digital Transformation Program Sponsor, a senior executive who stays involved, tracks progress, and takes action when needed. Influential executives should be included on the Program's Steering Committee. Ideally the sponsor would be complemented by having the same good experienced Digital Transformation Program Leader throughout the program's life. And a motivated, skilled Program Team. The team should have both business and IT skills.

A Digital Transformation Program is likely to run for several years. It's likely to lead to changes affecting many people throughout the organisation. It's important that executives recognise the need for a clearly defined and professionally run Organisational Change Management (OCM) activity in the program. They should ensure that an OCM activity is included in the Digital Transformation Program plan. It should include tasks such as increasing awareness of Digital Transformation; preparing and providing Digital Transformation training; developing new means of recognition; creating the communication plan; clarifying new job descriptions; communicating about changes; creating the training plan; and coaching.

References

1. Drucker P (2015) Innovation and entrepreneurship. Routledge. ISBN 978-1138168343
2. Stark J (2017) Product lifecycle management (volume 3): the executive summary. Springer. ISBN 978-3319722351

Chapter 20
Set the Objectives

20.1 Objectives of the Digital Transformation Program

The Digital Transformation of an organisation may affect many of its functions such as Portfolio Management, Product Management, Marketing, Sales, Research, Engineering, Regulatory, Logistics, Production, Service, Recycling, Finance, Quality, Business Process Management, Human Resources and Information Systems (IS).

One of the first steps towards successful Digital Transformation (DT) is to identify and clarify the objectives of the Digital Transformation Program.

The organisation's objectives for Digital Transformation are important. They express at a high level what's expected, what should be achieved. They're a statement of what the organisation wants to achieve, what it's aiming at. They will drive all the activities of the Digital Transformation Program. Everybody needs to know what they are.

20.2 Executive Involvement in Setting Objectives

Executive involvement in developing Digital Transformation objectives is essential. Only executives have the necessary strategic organisational understanding to define them. The Digital Transformation objectives need to be aligned with the organisation's objectives. Defining the objectives, defining where the organisation should be in the future, is an executive responsibility. The Digital Transformation Program looks at the long term, it's related to the future performance of the organisation, it may have high costs, and there's a need to prioritise activities. Only executives have the knowledge and experience to take the right decisions.

In this activity of setting objectives, executives have two main inputs. On one hand, they're aware of the organisation's vision and objectives for the future. On the other hand, they're aware of the many opportunities, potential benefits and issues in the scope of Digital Transformation.

Based on their knowledge of the organisation's objectives, and the many opportunities, issues and potential benefits in the scope of Digital Transformation, executives can make a first proposal for the objectives of the Digital Transformation Program.

20.3   Reasons for Setting Objectives

Figure 20.1 shows some important reasons for setting objectives.

20.4   Characteristics of Objectives

Objectives have some special characteristics (Fig. 20.2). These are worth remembering when setting objectives.

Objectives have to be defined in such a way that the targeted outcomes can be measured. Otherwise, no-one would know if they had been achieved. In the Digital Transformation context, one objective could be to increase revenues. But that sentence alone is vague. There also needs to be a definition of exactly how revenues are to be measured. And there needs to be a corresponding target, linked to a time period, such as an annual increase of 10% [1].

The process of developing the objectives helps get to common shared agreed objectives
Objectives show what a company is going to do to. Everyone will have clear targets in sight.
Objectives focus attention on what an organisation wants to achieve
Objectives express and communicate expected outcomes
An organisation's objectives for Digital Transformation drive all its DT activities
Objectives are a starting point for measuring progress
Objectives are a starting point for developing a strategy, and then a plan

Fig. 20.1 Reasons for setting objectives

Specific	defining exactly what the objective is in a particular situation
Quantified	otherwise objectives will be vague and of little value
Measurable	so they can be quantified and used to track progress
Documentable	so they can be written down as the basis for agreement and communication
Timely	the objective is to achieve something within a given time
Consistent	if there's more than one objective, they mustn't contradict each other

Fig. 20.2 Characteristics of objectives

20.5   Different Organisation, Different Program

Digital Transformation Programs are likely to be very different in different organisations because organisations are in very different situations (Fig. 20.3).

20.6   Different Organisations, Different Objectives

Because organisations are in very different situations, it's likely that the objectives set for the Digital Transformation Program in one organisation will be different from those in other organisations [2].

There are likely to be objectives addressing cost, quality and time (Fig. 20.4).

In addition to the objectives addressing cost, quality and time, there may also be some operational objectives for the Digital Transformation Program. These may be at the departmental level (Fig. 20.5).

20.7   Example of Objectives

Usually, there are several objectives for a Digital Transformation Program.

For example, based on input from executives, an organisation could develop the following annual objectives for Digital Transformation: Cost of Quality (COQ) reduction of 10%; cost reduction of 5%; Time to Market (TTM) reduction of 10%; introduction of at least one significantly innovative product; 5% growth in the value of the product and service portfolio.

Fig. 20.3 Different situations when starting a DT program

organisations offer different products and services
organisations have different positions in the supply chain
organisations are at different maturity levels of Digital Transformation
organisations have different Digital Transformation skills
organisations have different reasons for starting their Programs

Fig. 20.4 Examples of business objectives for the DT program

reduce costs	increase product revenues
reduce product cost	raise service revenues
reduce support costs	improve time to market
respond faster to changes	improve product quality

Fig. 20.5 Examples of operational objectives for the DT program

optimise resources	improve communication
ensure compliance	improve decision-taking
understand customers better	be more efficient
find information easily	use information better
support distributed teams	increase innovation

References

1. Rouillard L (2009) Goals and goal setting: achieve measurable results. Axzo Press. 978-1426018350
2. Stark J (2016) Product lifecycle management (volume 2): the devil is in the details. Springer. ISBN 978-3319244341

Chapter 21
Create the Vision

21.1 A Digital Transformation Vision

Development of a Vision of the targeted Digitally Transformed environment is often one of the first activities of the Program. A vision is a high-level conceptual description, a Big Picture of the future environment.

It's not enough for an organisation to say it's going to do Digital Transformation (DT). Unless it develops and communicates a clear vision of what it wants to achieve, nobody will know what the expectations are, and it's likely that everyone will see things differently.

A DT Vision is a high-level conceptual description of an organisation's activities at some future time. It's difficult to look further into the future than a few years. So it's appropriate to develop a vision of what the digitally transformed organisation will look like five years in the future.

A Digital Transformation Vision represents the best possible forecast of the desired future situation and activities. A DT Vision outlines the framework and major characteristics of the future position. It provides a Big Picture to guide people in the choices they have to make, when strategising and planning, concerning resources, priorities, capabilities, budgets, and the scope of activities.

There's a saying, "a ship without a destination doesn't make good speed". Without a DT Vision, people won't know what they should be working towards, so won't work effectively.

For some organisations, one step in the Digital Transformation Program will be to develop and communicate a DT Vision for the future environment. They'll do this to find out and understand where they should be going in the future, and to help everyone share this understanding.

J. Stark, *Digital Transformation of Industry*, Decision Engineering, https://doi.org/10.1007/978-3-030-41001-8_21

21.2   Need for a DT Vision

You may be wondering if it's really necessary for an organisation to have a DT vision. So here's an example to demonstrate its importance.

Imagine you ask me to book your next vacation. Sure, I say. But since you've told me nothing, I have no idea of what you're thinking of doing or where you're thinking of going. Put another way, I don't have a Big Picture of what you're dreaming of. Are you thinking of going to Lake Jocassee perhaps? No? So tell me, are you thinking of a vacation by the sea, in the mountains, perhaps a city break? I don't know. You may not have a great vacation unless you can share with me a high-level description of your vacation activities! Oh, OK, you'd like a vacation by the sea. Well, that gives me some idea. But what are you thinking about? Underwater, diving, lying on the beach, swimming, going out on a boat, fishing, or cruising in the Antarctic? And how do you want to get there? And are you going alone? And when do you want to go? Summer, winter?

It's a similar situation with an organisation saying it is going to digitally transform. Unless it develops and communicates a vision of what it wants to do, nobody will know what the expectations are, and it's likely that everyone will see things differently [1].

21.3   Reasons for a DT Vision

The vision is a good basis for communication about Digital Transformation between everybody involved, including executives, IS specialists, Customer Support Managers and product developers. The vision gives everyone in the organisation a clear agreed destination to work towards. The vision creates a framework against which decisions can be taken. It will guide people through strategy setting and planning, and help with the deployment of Digital Transformation. Without the vision, people wouldn't know where they're going, or what changes may be made. So they wouldn't be able to take reasoned decisions about what they should do in the future.

Organisations need a clear Digital Transformation Vision so they don't drift along, going wherever external forces are pushing them. People in the organisation need a clear agreed DT destination that everyone can work towards. A DT Vision for the organisation will enable all participants and decision-makers to have a clear, shared understanding of the objectives, scope and components of the organisation's Digital Transformation Program. It's a good basis for future progress. A DT Vision is a focal point for everybody in the organisation that says: "this is where we're going". The Vision is a useful basis for communication about DT between all those involved with Digital Transformation, such as executives, IS managers, Product Managers, sales associates, service staff, recycling managers and other stakeholders. It allows everybody to "work from the same book" and "sing from the same page" [2].

There are also other reasons for developing a Digital Transformation Vision. The vision is a Big Picture that will help people visualise the future. Developing the vision helps get consensus. It helps show what's missing, and makes sure that everything is included and nothing is forgotten.

Once the vision is developed, it should answer the questions that people will have had about DT. Questions such as: What's in Digital Transformation? What's not in Digital Transformation? What will Digital Transformation look like for us? How will it differ from today's world? What resources will be needed? What will be transformed?

Questions like these show how important it is to create and communicate a clear picture of the future DT environment. Otherwise many of these questions are likely to go unanswered. And it's unlikely the vision will be achieved.

21.4 Characteristics of a DT Vision

Digital Transformation Visions have some special characteristics (Fig. 21.1).

A Digital Transformation Vision will be organisation-specific. It's specific to an individual organisation because it depends on the circumstances and resources of the organisation, and on its particular environment. Without knowing a particular organisation in detail, it's not possible to say what its DT Vision should be. For example, the DT Vision of an airport would be expected to be different from that of a watch manufacturer.

A DT Vision is built on the assumption that the organisation wants to carry out its activities as effectively as possible. Organisations don't proactively set out to perform badly.

A Digital Transformation Vision must make sense to others. It has to be unambiguous and easily understandable. It must be believable and realistic, although it may appear to be at the limits of possibility. It must relate to the world of its readers, so that they can find a place for themselves within it. The Vision has to be as realistic and clear as possible. Otherwise it's not going to be useful.

A Digital Transformation Vision may sound as if it's ghostly and immaterial. However the DT Vision needs to be concrete, clear, complete, consistent and coherent. It needs to be understandable and meaningful to different types of people in the organisation. It needs to provide people at different positions in the organisation

| A DT Vision describes, within the bounds of its mission, an organisation's aspirations for the future |
| A DT Vision is a best estimate for the future. It's unlikely to correspond exactly to future reality. |
| A DT Vision must be documented |
| A DT Vision must be concrete, clear, complete, consistent, consensual and coherent |
| A DT Vision must be unambiguous, believable, realistic and easily understandable |
| A DT Vision must be organisation-specific |

Fig. 21.1 Characteristics of DT visions

with different levels of detail. A Vision that's incomplete isn't going to result in much progress.

The DT Vision helps communicate to many different people an overview of what Digital Transformation is, and what it will be, why it's important, and how it will be achieved. The Vision has to be communicated to everybody likely to be involved in the Digital Transformation Program's activities or impacted by them. It wouldn't make sense to have a Vision that's only accepted or understood by its inventor.

There has to be consensus about the Digital Transformation Vision. A shared Vision helps everybody to move forward towards an agreed target environment.

The purpose of the Vision is to document and communicate the focus, requirements, scope and components of DT. It communicates the fundamental "what's, why's and where's" of DT, and provides a framework against which decisions can be taken.

A DT Vision is the starting point for developing a DT Strategy, and for developing and implementing improvement plans. In the absence of a shared Vision, people won't have a common picture of the future to work towards, so plans and improvement initiatives might be unconnected or even in conflict.

The Vision is a best estimate for the future. It's the most likely Vision out of an infinite number of possible Visions for Digital Transformation. It's unlikely that the Vision will be the reality in five years. Most likely, new opportunities will arise over the five years and lead to a different reality. And, during the five-year period, the organisation will be in intermediate states on the way to the Digital Transformation Vision, rather than in the Vision state itself.

21.5   Documenting the DT Vision

For several reasons (Fig. 21.2), the Digital Transformation Vision should be documented in a formal report.

21.6   Position of a DT Vision

Once the Digital Transformation Vision has been agreed, a suitable Implementation Strategy has to be developed to achieve it. Once the Implementation Strategy has been defined, it's possible to start the planning of detailed implementation activities.

Fig. 21.2 Reasons for developing a formal Vision document

to distribute the Vision for review and correction
to distribute the Vision to increase awareness
to reduce the risk of the Vision being forgotten
to reduce the risk of misunderstanding
to reduce the risk of distortion in support of particular interpretations
to communicate the Vision

The resulting plans will address topics such as applications, modifications to the organisation's processes, information, and organisational structures. Individual projects will have to be identified, and their objectives, action steps, timing and financial requirements defined. The relative priorities of these projects will have to be understood. The projects will have to be organised in such a way that they result in the Digital Transformation Vision being achieved within the allowed overall budget and time scale.

When planning is complete, implementation can take place.

21.7 Executive Involvement with the DT Vision

Executives need to be involved in the visioning activity to ensure that the strategic focus and objectives of the organisation are taken into account. Upstream of the DT Vision are the organisation's mission, strategy and objectives. The DT Program needs to be aligned with the organisation's objectives. In most organisations, only executives have a sufficiently broad view of the mission, strategy and objectives to ensure the right vision is developed. The DT Vision is likely to miss some strategic components if executives are not involved in its development.

It's an executive role to take the lead and show the way.

A Digital Transformation Program is cross-functional, tough and risky. Without executive support and involvement it's likely to fail. If things aren't working out, it's likely that only executives will be able to put the Program back on track.

References

1. Rothauer D (2018) Vision and strategy: strategic thinking for creative and social entrepreneurs. Birkhauser. ISBN 978-3035614923
2. Stark J (2016) Product lifecycle management (volume 2): the devil is in the details. Springer. ISBN 978-3319244341

Chapter 22
Prepare the Governance

22.1 The Program's Governance

The Digital Transformation of an organisation may affect many of its functions such as Product Management, Marketing, Sales, Research, Engineering, Regulatory, Logistics, Service, Recycling, Finance, Quality, Human Resources and Information Systems (IS).

One of the first things to detail for the Digital Transformation Program is its governance, its governing structures and processes, its management, organisational and operating framework.

The governance is the overall set of policies, procedures, responsibilities, rights, roles, rules and structures that governs the program's activities. It includes descriptions of the main organisational entities in the program (for example, the Program Steering Committee and the Program Team) and the relationships between them. It describes the participants in the program (for example, the Program Sponsor, the Stakeholders, the Program Leader, and the Subject Matter Experts), their roles and the relationships between them.

22.2 Program Roles

The leading roles in the Digital Transformation (DT) Program will be the CEO, the DT Program Sponsor, the DT Program Steering Committee and the DT Program Leader [1].

22.3   CEO

CEOs aren't expected to be interested in all the low-level technical details of a DT Program. But they're likely to want to be involved with its high-level issues. For example: selecting or confirming the Program Sponsor; defining or confirming objectives; signing off the DT Program Charter; confirming the governance; confirming the Vision; and authorising the investment in the DT Program.

Then, throughout the DT Program, the role of the CEO could include: having regular meetings with the Program Sponsor; maintaining awareness and ensuring progress is being made; being informed when each phase is completed; authorising any major changes in scope; being available for discussions with the Program Sponsor on an as-required basis.

22.4   Digital Transformation Program Sponsor

The DT Program Sponsor, a senior executive, will have the most senior role in the Program. The sponsor will have ultimate authority, and will be responsible and accountable to the CEO for the delivery of the benefits planned for the Program.

Throughout the Digital Transformation Program, the sponsor will lead the Program and keep the focus on achieving its objectives. The sponsor will represent the DT Program, and communicate with stakeholders. In particular, they'll communicate and work with the DT Program Steering Committee. The sponsor will ensure the governance framework and mechanisms are documented, communicated and applied. They'll confirm the Program's objectives, ensure Program funding is available, agree the plan, and ensure Progress Indicators are in place. They'll report progress to the CEO. They'll provide direction, advice and support to the Digital Transformation Program Leader.

During the Initiation and Planning Stages of the Digital Transformation Program, the sponsor will set, or confirm, the Program objectives. They'll: appoint the DT Program Leader; ensure the Program governance is in place; agree the initial plan and expenditure; oversee development of the DT Program Charter; and ensure Program Progress Indicators are defined.

During the Execution Stage of the Digital Transformation Program, the sponsor will: ensure risks and issues are described, tracked, addressed and resolved; keep aware of progress; assist with major issues, problems, and conflicts; address proposed scope changes; chair meetings of the Steering Committee; review deliverables at the end of each phase of the Program; and approve progress from one Program phase to the next (Fig. 22.1). This person needs to be committed and available for the duration of the Program. The likely involvement of the Sponsor in the Digital Transformation Program is about 4 hours per week.

ensure that Program goals are achieved	take high-level decisions
assist with major issues, problems, and policy conflicts	chair the Steering Committee

Fig. 22.1 Typical activities of the DT Program Sponsor

resolve Program issues related to their department	review and approve scope changes
allocate resources to the Program from their department	approve end-of-phase Program deliverables
listen to progress reviews from the Program Leader	assist in securing funding
support the Program Sponsor in decision-taking	liaise with executive groups

Fig. 22.2 Typical activities of the DT Steering Committee

22.5   DT Program Steering Committee

The Digital Transformation Program Steering Committee will be made up of high-level executives from those parts of the organisation affected by Digital Transformation. It could include representatives from departments such as Marketing, Sales, R&D, Engineering, Manufacturing, Service, Recycling, IS, Quality and Finance.

The Digital Transformation Program Steering Committee will provide business and technical input for the Program. It will support the sponsor in the overall direction and management of the Program. It will advise on scope-related matters. Its members will: allocate resources to the Program from their parts of the organisation; support the Program in achieving its objectives; resolve any Program issues related to their part of the organisation; support the sponsor in decision-taking; and participate in approving progress from one Program phase to the next.

The Digital Transformation Program Steering Committee's exact activities and responsibilities are likely to differ from one Program to another (Fig. 22.2).

The Steering Committee is usually headed by the Program Sponsor. The likely involvement of a Steering Committee member in the Program is about 2 hours per week.

In view of the differing roles of Steering Committees in different programs, it's important to define the exact roles and responsibilities of an organisation's Digital Transformation Program's Steering Committee.

22.6   DT Program Leader

The Digital Transformation Program Leader will lead the DT Program with the aim of achieving the objectives of the DT Program. The DT Program Leader will be given authority by the sponsor to lead the Program on a day-to-day basis. They'll be responsible to the sponsor for on-time, on-budget achievement of the Program's goals. Typical activities are shown in Fig. 22.3.

prepare the Program plan	carry out high-level scheduling
manage the Program on a daily basis	manage Program resources
monitor progress against plans	select Program team members
communicate with top management	motivate the team
secure approval of deliverables	coach the team
receive direction from the Sponsor	report progress to the Steering Committee
manage risks and issues	get feedback from the Steering Committee
monitor contract compliance with externals	raise issues with the Steering Committee

Fig. 22.3   Typical activities of the DT Program Leader

Initially the Program Leader will need to understand the Program's goals, and its time, cost and other resource constraints. Then they'll need to prepare the Program governance and the DT Program Charter. The Program Leader will: clarify the activities required to achieve Program objectives; prepare the Program plan, detailing expected deliverables; identify potential Program Team members; build Program Teams; and assign Team members to tasks.

During the Execution Stage of the DT Program, the Program Leader will: manage the Program on a day-to-day basis; manage and lead the Program Team; onboard Team members; manage training activities for Team members; and motivate and coach Team members. The Program Leader will: hold Team meetings; monitor progress against plans; develop and maintain detailed plans; record and manage risks and issues; provide status reports to the Program Sponsor; prepare Steering Committee meetings; participate in Steering Committee meetings; and document deliverable and phase acceptance.

In large Digital Transformation Programs, there can be a huge load of program management and project management work. As a result, the Digital Transformation Program Leader role is often shared between a Program Director and a Program Manager, with the Program Manager reporting to the Program Director.

In such a situation, the Program Director usually addresses the more strategic issues and the communication with the Steering Committee. Meanwhile the Program Manager manages the day-to-day issues.

If the program management role in a Program is split, it's important to define the exact roles and responsibilities of the Program Director and the Program Manager.

22.7   Program Team, Program Team Member

Digital Transformation Program Teams come in various shapes and sizes. The Team may consist of people from just one functional area, although usually it consists of people from several functional areas. There may be just one team. Or there may be a Program Team made up of several teams. For example, there may be a Core Team, made up of the most involved people, and an Extended Team, made up of less-involved people [2].

understand the assigned task	plan detailed activities
execute the assigned task	report status, results and concerns

Fig. 22.4 Typical activities of a DT Program Team Member

provide knowledge of a specific subject	help people get facts straight
provide expertise in a specific subject	participate in data mapping
act as an authority on a specific subject	participate in defining tests
explain how tasks are currently executed	participate in workshops
review documents before communication	report weak points
participate in process modelling	participate in preparing training material
participate in use case development	participate in training activities

Fig. 22.5 Typical activities of a DT Subject Matter Expert

A Program Team Member executes tasks and produces project deliverables. They may be assigned full-time or part-time to the Program Team. They may carry out a wide range of activities (Fig. 22.4).

Some DT Program Team members may not have participated in a Program before. To avoid confusion and wasting time, it's important to define their roles and responsibilities.

22.8 Subject Matter Expert

A Subject Matter Expert (SME) is someone who has excellent knowledge and experience of their particular subject (or area or topic). An SME provides expertise on their subject as required by project activities (Fig. 22.5).

22.9 Stakeholders

In different Digital Transformation Programs, the term "stakeholder" may be used with different meanings.

At one extreme, the stakeholders are only the high-level executives whose departments and other domains in the organisation will be impacted by the Program.

At the other extreme, the stakeholders are all the people or groups which may impact, or may be impacted by, or may otherwise have an interest, in the Program.

In view of this wide spread of meanings, it's important to define the exact roles and responsibilities of a stakeholder in a particular organisation's Digital Transformation Program.

References

1. Stark J (2017) Product lifecycle management: the executive summary, vol 3. Springer. ISBN 978-3319722351
2. Stark J (2019) Product lifecycle management: 21st century paradigm for product realisation, vol 1. Springer. ISBN 978-3030288631

Chapter 23
Define the Strategy

23.1 Implementation Strategy

To achieve the Digital Transformation Program's Vision, the Implementation Strategy, or Change Strategy, has to be developed. The Implementation Strategy outlines how resources will be organised to achieve the transformation from the current situation to the future envisioned environment. It's the strategy to go from the as-is situation to the to-be situation.

23.2 Strategy

Originally the word "strategy" was used in a military context. The word itself comes from the Greek word for a General. In many dictionaries, "strategy" still has a primarily military definition. The following definition reflects the fact that strategy is no longer confined to the military environment. "A strategy is a general method for achieving specific objectives. It describes the essential resources and their amounts which are to be committed to achieving those objectives. It describes how resources will be organised, and the policies that will apply for the management and use of those resources".

Lessons can be learned from the application of different military strategies. These lessons learned are a useful input when developing strategies for Digital Transformation (DT).

Fig. 23.1 A small range of simple strategies

control of the seas	attack in overwhelming strength
control of the air	attack with overwhelming speed
control of a land region	destroy the enemy's will to fight
impregnable defence	blockade
divide the enemy's resources	siege
cut the enemy's communication lines	cut the enemy's supply lines

The range of military strategies is small (Fig. 23.1). These strategies all appear simplistic and are described in a few words. This is because strategies have to be simple. Otherwise, few people will be able to understand them. And even fewer will be able to implement them.

Over time, strategies change. As the environment and the resources change, strategies change. A strategy that may succeed at one time and in one place may be disastrous under other conditions.

It's sometimes said that Generals prepare to fight the last war. This can be seen in France in the First World War where the French generals' desire to attack stemmed from Napoleon's strategies applied in the previous century. But the conditions created by the development of machine guns and artillery meant that a defensive strategy was appropriate.

By the time of the Second World War, the value of defence had been understood by the French generals and the defensive Maginot line created. However, the resources available had changed again, and a strategy based on defence led to a French defeat a few weeks after the start of the German offensive [1].

The choice of strategy depends on the objectives and the surrounding environment. There's always a choice of possible strategies. No strategy is going to be right under all conditions. The only way to judge whether a strategy is right or wrong, is whether or not it results in the objectives being met [2].

23.3 Characteristics of Strategies

Strategies are specific to individual organisations because they depend on the particular circumstances and resources of the individual organisation, and on its particular environment.

Strategies change. Today's strategy describes how resources are used in today's environment. A strategy for the future shows how they'll be used in the future.

A strategy has to be documented and communicated to everybody likely to be involved in the future environment or impacted by it. It wouldn't make sense to have a strategy that nobody, apart from its developers, knows about, understands or approves.

A strategy shouldn't be changed frequently. It can take several years to implement a new strategy. And it can take several years for the effects of a new strategy to become apparent.

A good strategy shows how objectives will be achieved
A good strategy provides the best chance of achieving the Vision
A good strategy makes sure resources and capabilities are used to their best
A good strategy is a framework to take decisions and action
A good strategy is a communication tool informing everyone what's happening
A good strategy enables planning decisions to be taken in a coherent way
A good strategy includes Key Performance Indicators and targets to track progress
A good strategy helps everybody work towards the same targets

Fig. 23.2 Reasons why a good strategy is important

23.4 Reasons for a Good Strategy

There are several reasons why a good, well-defined, well-communicated strategy is important (Fig. 23.2).

23.5 Executive Involvement in Strategy Development

There are several reasons why executives need to be involved in strategy development [3].

The strategy answers the question, "How can we achieve the objectives?" It sets the direction for the company. Executives represent the company. They know what they want to achieve, and how they want to do it.

Executives are more aware of company issues than those at a lower level. Lower level people may not even be conversant with the issues faced at the top.

Participation by executives in strategy development enriches their insight and strengthens decisions. It builds executive ownership and commitment, and helps get consensus at the top level.

23.6 Description of the Current Situation

The Digital Transformation of an organisation may affect many of its functions such as Portfolio Management, Product Management, Marketing, Sales, Engineering, Production, Service, Recycling, Finance, Quality, Human Resources and Information Systems (IS).

Before developing the Implementation Strategy, the current situation of the environment to be digitally transformed needs to be understood. It's the starting point from which the Digital Transformation Program will progress. In another, much simpler, context, let's assume you want to go to "D". That may sound great, you know where you want to go. But how will everyone in your organisation get there? You haven't mentioned a starting point. Are you, and others, starting from "A", "B" or "C"?

Depending where you start from, you'll take a different road. It's the same principle in the Digital Transformation Program. You need to know from where you'll start.

It's important to have a clear description of the current situation. And a corresponding in-depth understanding. This understanding has to be based and built on documented facts, not on guesses, individual's opinions and vague conjectures. If the current situation isn't documented, there's no way of being 100% sure about it. If you're not sure what's there, and what the problems are, it's going to be difficult to know what should be improved, or how to improve it.

If you don't know the current situation, you may be missing key information that you need before making your proposal for the Implementation Strategy. There could be many things that work very well. You may not want to change them, because changing them might impair performance, not improve it. If you don't know the current situation, you may miss easy improvement opportunities. If you know what you do badly, you can make sure you don't do it again in the future, 'and don't propose the same wrong things for the future. And, to successfully implement change, you need to communicate it to, and convince, many people. You need to communicate a clear documented message. If you can't even explain how things are today, it's unlikely that anyone's going to believe your suggestions for getting to the future situation.

23.7 Analysis, Gap Analysis

To be able to develop the Implementation Strategy for the Digital Transformation Program, it's necessary to have a detailed understanding of both the current situation and the targeted future situation, the Vision [4].

Based on this understanding, the next activity is one of analysis: analysis of the current situation; and identification and analysis of the gaps between the current and future situations.

Gap Analysis is a technique often used to identify the steps that will need to be taken to transform from the current situation to the desired, future situation. It's based on a good understanding of both the current situation and the future situation. It's likely that there will be many differences in the descriptions of the two situations. The gaps, the differences, should be listed and described in a Gap Description Matrix.

With the gaps between the current and future situations identified and documented, the next step will be to look for ways to close them. Several ways should be found to eliminate each gap. They should be included in a Gap Elimination Matrix. By now, many activities that will help the organisation move forward to the future situation will have been described. Decisions about these activities will be considered in the development of the Implementation Strategy and the Implementation Plan.

The Implementation Strategy for the Digital Transformation Program will show how to get from the current use of resources to the future use of resources. There are likely to be many ways to do this, and the likelihood of finding the most appropriate at the first attempt is low. Several potential scenarios should be identified and documented. Each scenario will show a different way to reach the future situation. Each scenario should be described in detail. After the scenarios have been identified and described, they should be analysed. The strengths and weaknesses of each scenario should be described. Analysis of the scenarios leads to identification of the preferred Implementation Strategy.

The Implementation Strategy for the Digital Transformation Program is likely to be very different in different organisations. Their current situation is different, their future situation is different. The scope of activities considered is likely to be different. And there are many ways to get from the current situation to the future situation. So, it's to be expected that each organisation will create a different Implementation Strategy.

One example of an implementation strategy is the "Big Bang" strategy, with everything changing in Year n. Another example of an implementation strategy is the "Continuous Improvement" strategy which works with frequent small incremental improvements. A third example is a phased approach with a small set of changes being introduced in each phase [5].

References

1. Stark J (2016) Product lifecycle management: the devil is in the details, vol 2. Springer. ISBN 978-3319244341
2. Kourdi J (2015) Business strategy: a guide to effective decision-making. The Economist. ISBN 978-1610394765
3. Stark J (2017) Product lifecycle management: the executive summary, vol 3. Springer. ISBN 978-3319722351
4. Rodriguez M (2016) Lion leadership: teamwork, strategy, vision. Tribute Publishing. ISBN 978-0990600190
5. Stark J (2019) Product lifecycle management: the case studies, vol 4. Springer. ISBN 978-3030161330

Chapter 24
Plan and Schedule

24.1 Implementation Plan

The Digital Transformation (DT) of an organisation may affect many of its functions
such as Portfolio Management, Product Management, Marketing, Sales, Research,
Engineering, Regulatory, Logistics, Production, Service, Recycling, Finance, Quality, Business Process Management, Human Resources and Information Systems
(IS).

Digital Transformation may be enabled by technologies such as Analytics, apps,
Artificial Intelligence, automation, Big Data, Cloud computing, e-commerce websites, the Internet of Things, mobile, robotics, smart connected products, smart
phones, social, streaming, Virtual Reality and vlogs.

The Digital Transformation Implementation Strategy of an organisation addresses
the use of resources to change from the current environment to the future digitally
transformed environment. It's the starting point for developing and implementing
the DT Implementation Plan.

A plan is a detailed, organised method for doing something.

The DT Implementation Plan identifies the detailed activities and resources
needed to get to the future digitally transformed environment. It usually addresses a
multi-year timeframe. It addresses all the resources that will change. It will probably
contain many projects. Each of these is described in terms of objectives, action steps,
timing and financial requirements. The relative priorities of projects are described.

24.2 Strategy and Plan

Both a strategy and a plan describe how something will be done. But the strategy is
at a higher level than the plan. It outlines how objectives will be achieved. Whereas
a plan is the detailed way to do something.

The Implementation Strategy that has been developed may be for a phased approach, with each phase being 6 months long. It may show the organisation decided not to work with consultants. It may show the organisation decided to transform initially on just one site. But the Implementation Strategy doesn't show details such as: which projects are in each phase of the Digital Transformation Program; who will be doing what; and when they will be doing it. The development of the Implementation Plan will lead to those details.

24.3 Implementation Planning

Implementation planning helps turn strategy into action. It identifies the projects that make up the DT Program, the tasks needed to complete each project, and the people and other resources needed. Putting all the projects together, it shows how the objectives of the program will be achieved [1].

The Implementation Plan usually addresses a multi-year timeframe. It addresses all the resources that will be transformed. It contains many projects. Each of these is described in terms of objectives, action steps, timing and financial requirements. The relative priorities of projects are described. The plan makes clear the overall activities, resources and timelines of the organisation's Digital Transformation Program. It contains manageable and prioritised projects and sub-projects. It shows how the Vision will be achieved over the length of the program.

The Digital Transformation Implementation plan is more likely to be accepted if it includes some actions that will lead to short-term savings and other short-term benefits.

24.4 Implementation Schedule

The Digital Transformation Implementation Plan is complemented by the Program Schedule. The Program Schedule is a list or table showing the projects that make up the program. Each of these projects is defined and detailed in terms of tasks, intended start and end dates, deliverables, costs and participants. The relations between the projects are described. Each project in the table is usually broken down into sub-projects. In turn, these can be broken down into activities and tasks. Each of these is assigned to one or more people.

All organisations are different in many ways, so it's to be expected that a Digital Transformation Program Schedule will be specific to a particular organisation.

The schedule should address the long term and the short term [2].

For the long term, the schedule provides management with the information necessary to understand activities, resources and timelines. It should show how the implementation of Digital Transformation will be split into manageable phases (Fig. 24.1).

Phase Activity	Y1	Y2	Y3	Y4	Y5
Prepare Phase 1	■				
Execute Phase 1 activities	■				
Prepare Phase 2	■				
Execute Phase 2 activities	■	■			
Prepare Phase 3		■			
Execute Phase 3 activities			■		
Prepare Phase 4			■		
Execute Phase 4 activities				■	
Prepare Phase 5				■	
Execute Phase 5 activities					■

Fig. 24.1 Timing of DT program phases

Activity	M1	M2	M3	M4	M5	M6
Detail the plan for Phase 1 activities	■					
Manage the Phase 1 activities	■	■	■	■	■	■
Carry out activities related to mobile technology	■	■	■	■	■	
Carry out activities related to processes	■	■	■	■	■	
Carry out activities related to social technology	■	■	■	■	■	
Carry out activities related to the IoT	■	■	■	■	■	
Carry out Portfolio Management activities	■	■	■	■	■	
Finalise deliverables. Prepare report						■
Report Phase 1 activities						■

Fig. 24.2 Short-term DT program schedule

For the short-term, the schedule should show management which actions in the Digital Transformation Program will be taken initially, along with their timing (Fig. 24.2).

References

1. Lewis J (2010) Project planning, scheduling, and control: the ultimate hands-on guide to bringing projects in on time and on budget. McGraw-Hill. ISBN 978-0071746526
2. Stark J (2019) Product lifecycle management: 21st century paradigm for product realisation, vol 1. Springer. ISBN 978-3030288631

Chapter 25
Manage Organisational Change

25.1 Organisational Change Management

The Digital Transformation of an organisation may affect many of its functions such as Portfolio Management, Product Management, Marketing, Sales, Engineering, Logistics, Production, Service, Finance, Quality, Human Resources and Information Systems.

Organisational Change Management (OCM) is a structured approach, involving several Organisational Change activities, which accompanies and supports an organisation as it proactively changes from its existing organisational structure to a clearly-defined future structure. The objective of OCM is to successfully achieve this change [1].

Typical Organisational Change activities include aligning expectations of change, communicating about change, developing new recognition and reward systems, planning, training, coaching and mentoring. For successful change, the many change activities have to be planned and managed [2].

A Digital Transformation Program is likely to run for several years, leading to changes that will affect many people throughout the organisation. For the Digital Transformation Program to succeed, it's important to prepare for these changes with an Organisational Change Management (OCM) project in the overall program. This should include activities such as: preparing and providing training; planning tests for new applications; creating the communication plan; clarifying new job descriptions; and creating the training plan.

25.2 Relevance of OCM for DT

A Digital Transformation (DT) Program usually leads to many changes (Fig. 25.1).

The proposed changes will affect the way people work. For example, an improved New Product Development process will be executed by many people. They'll have

© The Editor(s) (if applicable) and The Author(s), under exclusive
license to Springer Nature Switzerland AG 2020
J. Stark, *Digital Transformation of Industry*, Decision Engineering,
https://doi.org/10.1007/978-3-030-41001-8_25

processes to be improved	new tasks to identify and define
old tasks to modify	people who will have to change
new documents to be used	cultural problems to address
new applications to bring in	organisational structures to change
data to be structured differently	new roles to be introduced
data to be used differently	new responsibilities to introduce

Fig. 25.1 Frequent changes in a DT Program

to understand and adapt to the changes. Similarly, a new data structure will be used by many people. They'll have to learn about the changes and work differently. New roles and responsibilities in the Engineering Change Management process will impact many people, and will change the way they work.

However, it's difficult for organisations and people to change, whether it's because of Digital Transformation, or because of another reason. Many people don't like to change. They have quite legitimate fears and concerns about change. They prefer things to stay as they are. However, if the changes don't occur, the objectives of the DT Program won't be met. In the absence of OCM, many DT Programs fail because the expected changes don't take place.

To avoid failure, it's important to identify and carry out activities to help change take place. Getting people to change is a major issue. Achieving success requires the application of "tools for change" such as learning, leadership, communication and the right reward systems.

Without an OCM project, the DT Program is unlikely to achieve its objectives. However, it's often difficult to get the people in a DT Program to understand and accept this. There may be several reasons for this. Many people in the Program will have technical backgrounds and be believers in hard facts. They may not be so keen on "soft" issues like OCM. Many people in the Program will be IS specialists, and assume that success will come from new applications, not from talking about change. Other IS specialists in the Program won't understand why "change gurus" are interfering with testing and Key User activities.

It can also be difficult to find experienced OCM specialists who can work in the DT Program. Digital Transformation is a relatively new subject and there's a shortage of "change practitioners" with experience of DT. And because the scope of Digital Transformation is very wide, it can be difficult for "change practitioners" without experience of DT to get their arms round it.

25.3 Resistance to Change

There are many reasons (Fig. 25.2) why it can be difficult for organisations and their people to change. Some of these have their source in the organisation's structure and way of working. Others have their source in its culture.

Fig. 25.2 Reasons for
resistance to change

attached to the old way of doing things	concerns about competence
feeling good about known routines	waiting for the storm to pass
seeing no benefits from change	not being informed
fear of job cuts	lack of trust
bloated with change	not being involved
fear of the unknown	fear of fads

In a large organisation, it's going to take a lot of time and effort to bring about change. Executives may feel the need to change, make the right decisions, and set the right targets. But unless the great mass of the employees change, then nothing's going to happen.

25.4 Benefits of OCM for DT

Digital Transformation may be enabled in an organisation by the introduction of digital technologies such as Analytics, apps, Artificial Intelligence, Big Data, Cloud computing, e-commerce, the Internet of Things, mobile, smart phone, and social.

As a result, many changes may be proposed in a typical Digital Transformation Program. However, it's difficult for organisations and people to change. But if they don't change, the objectives of the Digital Transformation Program won't be met.

Organisational Change Management accompanies and supports the overall organisational change. Its objectives are to make sure the change is successfully achieved and the objectives of the DT Program are met. OCM aims to bring benefits both to the organisation (Fig. 25.3) and to the people in the organisation.

OCM aims to support the individual employees who will be impacted by the targeted changes (Fig. 25.4). They will have to change from their current way of working to the future targeted way of working.

motivates everyone to achieve the targeted objectives	lowers the risks of change
plans involvement of the right people at the right time	anticipates challenges
maintains organisational effectiveness and efficiency	contains the costs of change
reduces the time needed to implement change	enables development of best practices for change
reduces the possibility of unsuccessful change	helps to align change resources

Fig. 25.3 OCM benefits for the organisation

smooths the transition from the old to the new	reduces stress
increases employee acceptance of the change	support for concerns regarding changes
ensures stakeholders understand and support the change	includes tasks for each person
improves cooperation, collaboration and communication	emphasises positive opportunities
acknowledges and addresses individual loss/gain	creates correct perception of the change

Fig. 25.4 OCM benefits for individuals

people will resist the changes	the Program's objectives won't be met
new IS applications will be underused	Program team members will be unhappy
there'll be confusion about what to do	supporters of the Program will be punished
executives will be unhappy	team members will leave the organisation
the Program Leader will be punished	Program team members will be punished

Fig. 25.5 Likely consequences of ignoring OCM in a DT Program

25.5   Results of Ignoring OCM in a DT Program

Many DT Programs fail. Some sources cite failure rates as high as 50%. Failure is rarely due to new technologies or changes to individual processes. Usually it's due to the way that these are implemented. It's often assumed that, without any support, people will switch overnight from the practices and tools they've been using for years to something completely new. In theory that may be realistic, but in practice it doesn't work, and the likely consequences can be foreseen (Fig. 25.5).

25.6   Generic Issues with Change

In a typical organisation, there will usually be many issues with Organisational Change in the Digital Transformation environment (Fig. 25.6).

There will usually also be many issues with the change activities that make up the OCM project of the DT Program (Fig. 25.7).

the required change isn't clearly defined	the required change isn't clearly communicated
the required change isn't clearly documented	ownership of the change isn't clear
the reason for the required change isn't clear	multiple versions of the same change
the objective of the required change isn't clear	no "change process for managing change"

Fig. 25.6 Generic issues with Organisational Change

not enough training	key people leave at short notice
not enough time to communicate	change of project scope
no process for OCM	change activities that don't add value
lack of visibility on the activities	unclear responsibilities
targets of change unclear	no agreement about change activities
too much detail	people don't understand the language
not enough detail	people don't understand the tools
inappropriate training	training given at the wrong time

Fig. 25.7 Generic issues with Organisational Change activities

Fig. 25.8 Examples of
activities related to
Organisational Change

prepare training for the new situation	align expectations of change
provide training for the new situation	plan tests for new processes
support individuals in new situations	clarify new responsibilities
provide awareness training about OCM	clarify new job descriptions
clarify the OCM approach and steps	develop new reward systems
develop new means of recognition	communicate about changes
plan tests for new applications	help restructuring activities
create the communication plan	develop an OCM Glossary
create the training plan	plan OCM activities
recognise achievements	prepare new roles
plan roll-out activities	mentor
create roll-out strategy	coach

25.7 Projects Related to OCM

In most DT Programs, there are many activities addressing change (Fig. 25.8)
Depending on the Program, some of these may run independently. Some may run in
parallel, or overlap. Many will be linked to other projects in the DT Program.

It's likely that many of the people in the DT Program will know little about OCM.
Coming from different backgrounds they may have very different understandings
of OCM. So it's often helpful to develop a glossary that gives short definitions of
the various terms and techniques used in OCM. This should lead to a common
understanding of the subjects to be addressed in OCM activities.

Similarly, some of the people in the OCM Team may know little about OCM.
Many of the team members may never have participated in change activities. It may
be difficult for them to know how best to plan their activities and go about their work.
As a result, some training about organisational change and Organisational Change
Management will be useful.

25.8 Executive Role with OCM

Digital Transformation is an executive issue. Executives define the objectives of
Digital Transformation. They should also define the objective of Organisational
Change.

Executives are responsible for providing the appropriate resources for the OCM
activity. They need to put in place the right people to work on Organisational Change
Management. Once the objectives of the Digital Transformation Program have been
defined, and the scope and size of the changes are clear, they should select the OCM
Team leader.

Executive participation in OCM is needed to ensure that Organisational Change
issues are taken into account in the DT Program. Due to the organisation-wide scope
of Digital Transformation, leadership has to come from the C-level.

Communication is a key issue in a change project. Executives should ensure good
communication of the needs and reasons for change.

Executives should convince middle managers of the need for change. Without the support of middle managers, the Digital Transformation Program is unlikely to succeed.

References

1. Kotter J (2012) Leading change. Harvard Business School Press. 978-1422186435
2. Stark J (2019) Product lifecycle management: 21st century paradigm for product realisation, vol 1. Springer. ISBN 978-3030288631

Chapter 26
Justify the Program, Plan the Investment

26.1 Justification

There are different ways of justifying the cost of a Digital Transformation Program. They range from having a general belief that the program will be beneficial through to carrying out a detailed financial justification.

An organisation may have a general belief that a Digital Transformation Program is likely to be beneficial because it will eliminate weak points.

Alternatively an organisation may have a general belief that a Digital Transformation Program is likely to be beneficial because Digital Transformation will allow it to seize opportunities.

In both cases, it's likely that the organisation will want to estimate the costs and benefits of the Digital Transformation Program and plan the related investments [1].

26.2 Opportunities and Weak Point Elimination

One of the first activities in a Digital Transformation Program is often to brainstorm the opportunities and pain points in its scope. This can be done with groups of 10–15 people. The groups should include people working in different parts of the organisation and at different levels. Most of these people will see opportunities and pain points time and time again in their daily work.

Most of them will be aware of at least ten opportunities and ten pain points. For example, people might mention issues such as: slow customer service; poor change management; limited customer choice; ineffective fixes; departmental information silos; islands of automation; product recalls; multiple definitions of a data element; incorrectly structured data; poor support of customers; missing applications; unclear business processes; ineffective application interfaces; missing skills; and lack of training.

J. Stark, *Digital Transformation of Industry*, Decision Engineering, https://doi.org/10.1007/978-3-030-41001-8_26

People might mention opportunities such as: reduced time to market; reduced delivery time; better and faster understanding of customer requirements; faster response to customers; improved decision-taking; improved collaboration; improved communication; improved compliance; improved risk management; fewer equipment breakdowns; reduced out-of-stock situations; better information for customers; simplified purchase activity for customers; better knowledge of products in the field; better sharing of knowledge; better adherence to standards and more in-service upgrades. They might also mention opportunities through adoption of approaches such as process mapping and Lean techniques.

Brainstorming can also be carried out with customers to try to find out how they view the organisation. This may lead to other opportunities and pain points.

Customer feedback is an important input for any decision about a Digital Transformation Program. The organisation might find out, for example, that customers thought it: didn't provide clear information about its services; had an unclear time-consuming order process; gave unreliable delivery dates; didn't offer good support; sent too many e-mails; ignored customer requirements; was too slow; charged too much for customisation; and had an unreliable website.

26.3 Program Costs

Having identified weak points and opportunities, the next activity may be to identify, describe and estimate the costs of the new technologies that are believed may help meet the objectives of the Digital Transformation Program. These technologies could include: Big Data, Analytics, Artificial Intelligence, Machine Learning, blockchain, Cloud computing, connected smart products, e-commerce, the Internet of Things, mobile technology, 3D Printing, robotics, social technology, Augmented Reality, Virtual Reality, immersive technology, automation, smart phone, Knowledge Management, and video streaming.

Probably some members of the organisation will know about many of the technologies. They could find out about others by looking on the Web and, for example, reading conference articles and technology vendor documents. It shouldn't take long for the organisation to list the new technologies, and describe them briefly in a report.

Identification of the costs associated with a technology introduction project is usually not too difficult. Costs are generally divided into initial investment costs (Year 0) and on-going costs (Year 1, and following years). Typical costs are shown in Fig. 26.1 [2].

26.4 Program Benefits

Two types of benefits are possible. Those that result in a reduction in costs. Those that result in increased revenues.

investment in software	software maintenance costs
technology selection costs	travel costs
investment in infrastructure	communication charges
costs for customising software	loading data in new systems
system management and operations	cleaning data
development of new working procedures	development of interfaces
on-going training and education	modification of existing procedures
participation in conferences and user groups	on-going consultancy
cost of feasibility studies	prototyping costs

Fig. 26.1 Typical sources of costs in a Digital Transformation Program

increase the number of customers	increase the percentage of customers re-ordering
increase the price paid by customers	increase the frequency with which customers buy
increase the number of products a customer buys	increase the range of services that customers buy

Fig. 26.2 Some ways in which DT can increase revenues

reduce the number of product developers	reduce costs of information systems
reduce the number of support staff	reduce the cost of quality
reduce the cost of materials	reduce energy consumption
reduce warranty and rework costs	reduce inventory costs

Fig. 26.3 Benefits that result in a reduction in costs

There are many ways in which Digital Transformation can increase revenues. Some examples are given in Fig. 26.2.

A reduction in costs is another potential benefit of Digital Transformation. This can be achieved in many ways (Fig. 26.3).

The proposed benefits of the new technologies and approaches should be documented. This helps ensure there's a good understanding of all the new technologies and approaches. It also helps avoid selecting a particular technology or approach without being aware of the benefits of other technologies and approaches. It also makes sure that the potential benefits are documented and communicated. At the top level, the benefits are often to reduce time, reduce cost, and improve quality.

The benefits described in previous activities may have seemed theoretical and vague to some people in the organisation. However there's now a change of focus. The theoretical, generic benefits proposed for the technologies and approaches are translated into specific benefits for the organisation's customers.

The intention is, for example, to identify that "if we do A, then the customer would benefit in three ways: B, C and D". Customer benefits are listed and documented in a report. The organisation's business process maps are revised in the light of the expanded knowledge and understanding. Workshops are held to identify potential improvement areas. For example, capturing and using information earlier, interacting faster with customers, removing tasks that add no value, adding tasks that increase customer satisfaction, using a new approach or technology, integrating applications, merging documents, and removing bottlenecks.

Potential improvements are listed and described, showing where they're positioned on the business process maps, explaining the benefits, describing the impact, and describing what's involved. Examples of potential improvements could include: showing the customer what the product/service is and where it can be acquired; showing the customer a video of how to assemble the organisation's product; showing the customer a video of how to fill in a form; showing the customer a virtual representation of something they'll only see later in finished form; informing the customer about the expected arrival time of the product/service; showing reliability of data on expected arrival time; showing how there's reduced effort for the customer; showing how the customer will save time; showing options, such as low-cost delivery in two days, or a same-day but higher-cost delivery; and enabling the customer to receive information anytime, anywhere with a smart phone app.

26.5 Financial Justification, NPV and ROI

In some organisations, when a Digital Transformation Program proposal is presented to executives, it will need to contain a financial justification that shows the required investment and running costs, the expected benefits, the expected return, the risks associated with the investment, and the effect of the investment on other areas of the organisation. Without such a justification, executives will be unable to decide either if the program is worthwhile, or if it's a better choice for investment than other projects and programs.

Several methods are used to express in understandable and comparable terms the profitability of different projects. These include: Accounting Rate of Return (ARR); Payback time; Net Present Value (NPV); Discounted Cash Flow Return On Investment (ROI).

The ARR is obtained by expressing, as a percentage, the ratio of the accounting income (including depreciation) generated by the project to the total investment. This is a quick calculation, but its inclusion of depreciation means that it's not all that useful from the project cash-flow point of view.

Payback time is the time required for a project's revenues to equal the cash outlay. If the investment in the project is \$1,000,000 and the annual revenue is \$400,000, then the payback time will be 2.5 years. Payback is a quick and dirty calculation. It doesn't take account of the time value of money, or of revenues occurring after the payback period. It gives a quick, approximate feeling for a project's viability.

The Net Present Value (NPV) of a project at any given time is calculated by subtracting, from the investment, the sum of the discounted cash flows up to that time.

$$NPV = -I + \sum_{t=1,n} \frac{(Rt - Ct)}{(1 + DR) ** t}$$

where I = investment in Year 0; n = project lifetime in years; Rt = revenue in Year t; Ct = costs in Year t; DR = discount rate.

For example, if I = \$1000, DR = 20%, n = 2 years, R1 = \$850, R2 = \$1050, C1 = \$250, C2 = \$300 then NPV = $-1000 + 500 + 520.83 = \$20.83$.

When the NPV is positive, the discounted cash flows are greater than the initial investment, so the project is earning more than the discount rate in use (in this case more than 20%). If the NPV turns out to be negative, then the project is making less than the discount rate in use.

In some cases, particularly when making comparisons between several competing projects, it may be enough to know the NPV. In other cases, it may be more useful to know the exact return of the project (also known as the Internal Rate of Return, or the ROI). This is the discount rate that corresponds to the Net Present Value of the project being equal to zero, i.e. the investment is exactly equal to the sum of the discounted cash flows. Once again, a simple calculation is all that's needed. Putting NPV = 0 in the above equation, and solving for DR,

$$I = \sum_{t=1,n} \frac{(Rt - Ct)}{(1 + DR) * *t}$$

so DR = 21.6, i.e. the rate of return of the project is 21.6%.

Even when all the cash flows of a project have been identified, and the ROI calculated, questions of the type "but what if...?" still remain. Sensitivity analysis and risk analysis try to answer them. Sensitivity analysis identifies the items that critically affect the project calculations. Risk analysis provides a range of possible values for the outcome of the project, rather than a single value.

Sensitivity analysis is used to look at each cash flow item individually, and answer the question "what would be the effect on the project's ROI if all other items have been estimated correctly, but this particular one over-estimated or under-estimated by x percent?" Each item can be checked in this way, and usually it's found that there are a few items that have much more influence on the ROI than the others. For example, a 10% variation in one item may lead to a 10% change in the ROI, whereas the same variation in another item may only lead to a 1% change. When the analysis has been carried out, the items that have the most influence (are the most sensitive) should be re-examined to make sure that they're based on as reliable and accurate information as possible.

Risk analysis is carried out to estimate the probability that the ROI will be met. One way of doing this is to assign probabilities to expected values for each cash flow item. Thus, instead of assuming that the value for a particular cash flow item will be \$7000, it could be estimated that there's a 5% probability that it will be \$5000, 10% that it will be \$6000, 70% that it will be \$7000, 10% that it will be \$8000, and 5% that it will be \$9000. Similar probabilities could be calculated for the other items. The ROI would then be calculated as a function of these probabilities. The result would show the range of values for the ROI, and the probability associated with each value.

References

1. Leimberg R, Robinson T, Johnson R (2017) The tools and techniques of investment planning. The National Underwriter Company. ISBN 978-1945424441
2. Stark J (2016) Product lifecycle management: the devil is in the details, vol 2. Springer. ISBN 978-3319244341

Chapter 27
Run the Program

27.1 Frequent Steps

From my experience with many organisations' programs, it's clear that there are some activities that appear time and time again (Fig. 27.1). For example, activities to define objectives, carry out a Feasibility Study, plan the Digital Transformation (DT) program, define an Organisational Change Management (OCM) sub-project, launch the program, define the Big Picture of the targeted future situation; understand digital technologies, understand the starting position (the current situation), develop the Implementation Strategy and the Implementation Plan, implement the plan, select digital technologies, and review program progress [1].

Some organisations will carry out all the activities. Others won't do some of them. One organisation will do the activities in one order, other organisations will do them in different orders. Different organisations will put these activities together in different ways as they build their Digital Transformation programs. The contents and structure of the resulting Program often depend on what the organisation has already achieved.

With so many differences between the needs and situations in different companies, it's not surprising that the steps taken in the resulting DT Programs will be very different. The resulting Programs will all be multi-step but the details will be different in different companies. Some examples are shown in Figs. 27.2 and 27.3.

carry out a Feasibility Study	develop the DT Implementation Strategy
understand the Current Situation	develop the DT Implementation Plan
develop the DT Vision	build a Financial Justification of the Program
develop the DT Strategy	develop the DT Program Charter

Fig. 27.1 Frequent steps in a Digital Transformation Program

© The Editor(s) (if applicable) and The Author(s), under exclusive
license to Springer Nature Switzerland AG 2020
J. Stark, *Digital Transformation of Industry*, Decision Engineering,
https://doi.org/10.1007/978-3-030-41001-8_27

Path 1	Path 2
Launch the DT Program	Launch the DT Program
Understand the Objectives of DT	Carry out a Feasibility Study
Understand the Current Situation	Understand the Current Situation
Understand the Future Situation	Develop the DT Strategy
Develop the DT Strategy	Develop the Implementation Strategy
Develop the DT Implementation Strategy	Develop the DT Implementation Plan
Develop the DT Implementation Plan	Implement the Plan
Implement the Plan	

Fig. 27.2 Two paths to DT implementation

Path 3
Carry out a DT Audit
Review DT Vision/Strategy
Review Audit Results
Review Implementation Strategy
Adjust the Implementation Plan
Implement the Plan

Fig. 27.3 A third path to DT implementation

27.2   Feasibility Study

Once executives have proposed objectives for the Digital Transformation program, there will be many ideas about what to do next. There will be many different ideas about the activities that will make up the program, and about the order in which they should be carried out. A Feasibility Study is a good way to determine the viability of these ideas, showing whether they're technically and economically justifiable. A Feasibility Study can be a good, low-cost way to get a better understanding of the ideas that are proposed [2].

Among the ideas to be reviewed could be: Alternative A: Don't do anything; Alternative B: Run a departmental Digital Transformation activity; Alternative C: Take a cross-functional approach to Digital Transformation; and Alternative D: Execute a strategic organisation-wide Digital Transformation Program.

The activities involved in reviewing each of these alternatives are similar (Fig. 27.4).

The first activity is to document the objectives and scope of the alternative. The next activity is to identify and quantify the benefits of achieving the objectives under this alternative, and estimate their financial value.

document the objectives and the scope of the option
identify the benefits of achieving the objectives, and estimate their financial value
identify the activities and effort required to achieve the objectives, and estimate their cost
create the business case
create an outline plan for implementation of the activities identified

Fig. 27.4 Activities for each of the alternatives

Revenue increases can be achieved in many areas such as through: an increased range of products; an increased range of services; increased sales of existing products; increased sales of new products; increased service revenues. If there's an opportunity with Digital Transformation to increase revenues in any of these ways, then people from Marketing, Sales, Service and Finance should be able to estimate it.

Another way to understand the benefits of Digital Transformation is to focus on the cost reductions it can provide. There are many areas in which costs can be reduced, such as: delivery costs; sales costs; energy costs; rework costs; development costs; documentation costs; recall costs; material costs; warranty costs; and service costs. If there's an opportunity with Digital Transformation to reduce costs in any of these areas, then people from Quality, Engineering, Manufacturing, Service and Finance should be able to estimate it.

Another activity for each alternative is to identify and quantify the tasks, participants and effort required to achieve the objectives with this alternative, and estimate their cost. And then, develop the business case, showing the approximate costs, financial benefits, and Return on Investment.

As a result of the Feasibility Study, the organisation will have a better understanding of the viability of the alternatives. It's likely that one of them will be more appealing than the others. This alternative could then be taken forward and further detailed.

27.3 The Program Charter

The development of a Charter for the program is another important activity in an organisation's Digital Transformation Program. The DT Charter (Fig. 27.5) describes and authorises the program. It's a formal document that outlines the reasons and objectives for the program, its cost and benefits, and the resources that will be involved. It makes clear what the program will achieve. Everything that anyone needs to know about the program is in the Charter, in one document. The Charter creates a common understanding, authorises the program, clarifies roles and responsibilities, and shows stakeholder commitment. The Charter is the authorising document for the program. It's created and signed off before Program Execution activities start.

27.4 Program Progress Reporting

With so much to do in a Digital Transformation Program, team members sometimes forget its objectives. They concentrate so much on short-term tasks, requiring completion in a few days or weeks that the overall objectives disappear over the horizon. However, executives are not so interested in the results of day-to-day tasks. They want to see progress towards the targeted objectives. To keep stakeholders informed,

DT Program Charter	
Table of Contents	
1 Introduction	6 Duration of the Program
1.1 Purpose of this Program Charter	6.1 DT Roadmap and Major Milestones
	6.2 Timeline
2 Executive Overview of the Program	6.3 Plan and Schedule for Year 1
3 Justification for the Program	7 Budget for the Program
3.1 Business Objectives	7.1 Estimate
3.2 Business Impact	7.2 Funding
3.3 Strategic Positioning	7.3 Budget for Year 1
4 Scope of the Program	8 Organisation of the Program
4.1 Objectives	8.1 Roles and Responsibilities
4.2 Business Requirements	8.2 Stakeholders (Internal and External)
4.3 Major Deliverables	
4.4 Boundaries	9 Approval of the Program Charter
5. Assumptions and Risks	10 Appendices
5.1 Assumptions	A Referenced Documents
5.2 Risks	B Glossary
5.3 Dependencies	

Fig. 27.5 Example of the contents of the DT Program Charter

#		Target	To Date
1	Rate of introduction of new products	+100%	+25%
2	Revenues from extended product life	+25%	+4%
3	Costs due to recalls, failures, liabilities	-75%	-5%
4	Development time for new products	-50%	-5%
5	Cost of materials and energy	-25%	-4%

Fig. 27.6 Example of progress towards targets

the team should develop and apply procedures to capture, at regular intervals, the data from which such information can be prepared (Fig. 27.6).

27.5 DT Program Status Review

12–18 months into the Digital Transformation Program, it's often useful to review the status and the progress of the program. The intention is to take the opportunity to stand back from everyday program activities, get a clear picture of the situation, and identify progress towards the Big Picture of Digital Transformation.

One benefit of a status review is to prevent any waning of interest in the Digital Transformation Program. Such waning of interest can lead to serious effects (Fig. 27.7). In extreme cases it can lead to the Program slowly grinding to a halt.

Another benefit of a status review is to keep the focus on the objectives of the Digital Transformation Program. Unless the focus is maintained on the expected

the DT Steering Committee stops steering	executives lose interest
users complain about applications	users carp about bureaucracy
middle management gets defensive	Key Users stop participating
users complain about project management	users complain about slow progress
service providers don't take responsibility	the DT Program Team is left with all the problems

Fig. 27.7 Effects of waning interest in the DT Program

an independent assessment of DT Program progress	A high level of acceptance of results by executives
an opportunity to communicate progress	an opportunity to focus again on DT Program objectives
expert advice concerning next steps	support and communication of next steps

Fig. 27.8 Advantages of a DT Program Audit by an external expert

benefits of a program, it's not likely that they'll be attained. There are usually plenty of ways for a program to change course, and zigzag off in the wrong direction.

A third benefit of a status review is to keep the program visible on executive control screens. It's only too easy for executives to agree to start the program, and then, faced with other important tasks throughout the organisation, forget about it, assuming it's making good progress.

Another benefit is that a Digital Transformation Status Review gives an opportunity to communicate progress towards targets, and maintain program momentum. The resulting Digital Transformation Status Review Report could include sections addressing: progress against plan; progress towards benefits; any risks and issues in the program; improvement suggestions; actual costs, planned costs; and next steps. This report will be a valuable resource to help with planning and moving forward.

27.6   DT Program Audit

It's often helpful to get an external expert to carry out an audit of the Digital Transformation Program. This can have several advantages (Fig. 27.8).

In many cases, the audit will show that the Program is on the right track, and making good progress. It may also show that a few improvements and adjustments may need to be made.

27.7   Digital Transformation Program Closure

Program Closure is an important step in the Digital Transformation program. Without it, the program would continue to consume resources indefinitely.

Typical closure activities are to: close out and document final tasks, deliverables and issues; complete and archive program documents; carry out a final program review; capture and document Lessons Learned; transition from the program mode to an everyday use mode; get sponsor and other stakeholder sign-off; and celebrate success.

References

1. Stark J (2019) Product lifecycle management: 21st century paradigm for product realisation, vol 1. Springer. ISBN 978-3030288631
2. Stevens R, Sherwood P (1982) How to prepare a feasibility study: a step-by-step guide including three model studies. Prentice Hall Direct. 978-0134292588

Printed in the United States
By Bookmasters

Printed in the United States
By Bookmasters